C. Sero

6/96

Paul D. Alb

Process
Drying Practice

Other McGraw-Hill Chemical Engineering Books of Interest

AUSTIN ET AL. • *Shreve's Chemical Process Industries*

CHOPEY, HICKS • *Handbook of Chemical Engineering Calculations*

DEAN • *Lange's Handbook of Chemistry*

DILLON • *Corrosion Control in the Chemical Process Industries*

FREEMAN • *Hazardous Waste Minimization*

FREEMAN • *Standard Handbook of Hazardous Waste Treatment and Disposal*

GRANT, GRANT • *Grant & Hackh's Chemical Dictionary*

KISTER • *Distillation Operation*

MILLER • *Flow Measurement Engineering Handbook*

PERRY, GREEN • *Perry's Chemical Engineers' Handbook*

REID ET AL. • *Properties of Gases and Liquids*

REIST • *Introduction to Aerosol Science*

RYANS, ROPER • *Process Vacuum System Design and Operation*

SANDLER, LUCKIEWICZ • *Practical Process Engineering*

SATTERFIELD • *Heterogeneous Catalysis in Industrial Practice*

SCHWEITZER • *Handbook of Separation Techniques for Chemical Engineers*

SHINSKEY • *Process Control Systems*

SHUGAR, DEAN • *The Chemist's Ready Reference Handbook*

SHUGAR, BALLINGER • *The Chemical Technician's Ready Reference Handbook*

SMITH, VAN LAAN • *Piping and Pipe Support Systems*

STOCK • *AI in Process Control*

TATTERSON • *Fluid Mixing and Gas Dispersion in Agitated Tanks*

YOKELL • *A Working Guide to Shell-and-Tube Exchangers*

Process Drying Practice

Edward M. Cook

Harman D. DuMont

Process Drying Practice is now sold by
the authors. Price $38.00 + 4.00 S/H
Send P.O. or check payable to
ESCO (Energy Saving Consultants)
1215 SW 22 Av., Boynton Beach, FL 33426

E. M. Cook   407-734-3504 (Fax 407-738-9363)
H.D. DuMont 908-534-6469 (Fax 908-526-1310)

McGraw-Hill, Inc.

New York St. Louis San Francisco Auckland Bogotá
Caracas Hamburg Lisbon London Madrid
Mexico Milan Montreal New Delhi Paris
San Juan São Paulo Singapore
Sydney Tokyo Toronto

Library of Congress Cataloging-in-Publication Data

Cook, Edward M. (Edward Moore), date
  Process drying practice / Edward M. Cook, Harman D. DuMont.
    p. cm.
  Includes index.
  ISBN 0-07-012462-0
  1. Drying. I. DuMont, Harman D. II. Title
  TP363.C657 1991
  660'.28426—dc20   90-22161
                                                         CIP

Second printing.

1 2 3 4 5 6 7 8 9 0 DOC/DOC 9 7 6 5 4 3 2 1

ISBN 0-07-012462-0

The sponsoring editor for this book was Gail F. Nalven, the editing supervisor was Ingeborg M. Stochmal, and the production supervisor was Suzanne W. Babeuf. This book was set in Century Schoolbook. It was composed by McGraw-Hill's Professional Book Group composition unit.

Printed and bound by R. R. Donnelley & Sons Company.

To all who have labored to improve drying technology

Contents

Preface

Drying is more art than science, they say, and this belief persists with good reason. In most industrial dryers, solid, liquid, and vapor states interact vigorously, and the taking of in-progress measurements is difficult or impossible. Unlike heat exchangers, and most other process equipment, dryers cannot be designed by fitting flow rates and properties into equations derived to suit the equipment's configuration. Instead, designs are based on tests. The result has been an accumulation of know-how by individuals, rather than a spread of general drying knowledge throughout the industry.

Most of the published works on drying report on fundamental studies of the internal conditions of heat and mass transfer, that is, the microscopic behavior of materials while drying. These help in understanding the process, but have limited usefulness in selecting, designing, and operating dryers. Instead, it is the external conditions of temperature, flow rates, and design features, together with experience, that meet industry's needs. This text focuses on these practical considerations.

Our aim is to record our know-how and experience in order to help others design, build, select, operate, and optimize drying systems. We hope to make others more knowledgeable in this puzzling business and guide them out of some of the common pitfalls.

The techniques and information offered either have never been published or are a more complete development of articles by the authors. Since dryer descriptions and theory are readily available, only what is essential for understanding each topic has been included.

The variety of products that have to be dried has evolved an abundance of dryer types, some for quite limited uses—no book could treat all of them in detail. The material here applies in some measure to all dryers, but specific emphasis is on those direct and indirect dryers most widely used for particulate materials—spray, flash, fluid bed, conveyor, tray, rotary, disc, paddle, screw flight, and drum.

Manufacturers of dryers are good sources of practical information, but many are firms with a niche, and are relatively small. Even most larger firms are knowledgeable in only one or two of the dozen or more basic dryer types. It is understandable that some of their information is biased and some is withheld for competitive reasons. They are limited, for these reasons, in the breadth of knowledge and assistance they offer. Recently their ranks have been thinned out. Most serious has been the loss of some product testing capabilities. Selecting dryers is thus more difficult. Not as tangible, but perhaps as serious, is a lessened incentive to innovate and to price competitively.

In an old technology such as drying, major changes come slowly. But some new approaches have been developed over the past few years to reduce energy use and increase dryer productivity. These techniques have been refined and successfully applied on a variety of drying systems, and they can also be used to improve test methods and results.

For these reasons, this seems a suitable time to disclose the practical side of drying technology. Included are some of the obscure working tools of the industry: scaleup, exposure times, testing methods, relations between drying conditions and product properties, complexities of psychrometric charts, calculation methods for any solvent in any gas, adversities in start-ups, finding leaks, and other items translated from drying folklore.

Chapters 1 through 12 are concerned largely with drying that removes water. The first outlines the internal and external drying conditions—the basic theories. Chapter 2 covers the main types of drying systems and their capabilities. The two chapters that follow describe specific dryers, separating them by heating methods into indirect and direct types.

Next are given the equations needed for the functional design of dryers—the airflow, heat load, and saturation conditions. It is shown how the equations can be used manually or in computer programs. Chapter 6 presents psychrometric charts made just for drying and tells how they are drawn and used.

Then two chapters detail the techniques and instruments needed to measure operating conditions, and how to make computer analyses of these data to improve energy use and productivity. Chapter 9 tells how dryers are selected by product-testing methods and how feed and product properties affect results. The advantages and disadvantages of buying used dryers are weighed in Chapter 10, followed by chapters giving advice and precautions for start-up and troubleshooting.

The realm of drying from nonaqueous solvents is covered in Chapter 13, and some unique estimating charts are presented. The final chapter addresses the reality of four actual installations, and how difficulties were overcome in the steps of testing through operation.

The appendixes list data needed for manual and computer calculations and conversion factors for both individual and combined units to and from U.S. customary and SI units. Throughout the text U.S. units are used, with SI units added in parentheses wherever practical. Next follows a glossary of terms. Finally, the references are for the most part those that either provided essential data or give useful background information. While some texts have proven invaluable, much more information and know-how has come from personal contacts and experience.

Acknowledgments

Special thanks are offered to William L. Root III of Komline-Sanderson Engineering Corporation, who provided much background and specific material on indirect dryers. Thanks are also due to Frank W. Dittman of the Center for Plastics Recycling Research for his review of drying theory and Chapter 1, also to Donald W. Belcher of Belcher Engineering, Inc., for reviewing Chapter 5, to Raymond H. Bosworth of Damrow Company for reviewing Chapter 6 and its charts, and to Ronald H. Gale of Universal Process Equipment Company for providing an inside look at the used-equipment market.

Much of this book stems from the authors' backgrounds, but some has also come from contacts over a period of years with knowledgeable individuals in the drying industry. Those who helped recently with their specific expertise include David W. Dahlstrom of ABB Raymond, John J. Walsh and Robert A. Nichols of Bepex Corporation, Fred Keill of Proctor & Schwartz, Inc., and Fred A. Aiken of Swenson Process Equipment, Inc.

Edward M. Cook
Harman D. DuMont

Chapter 1

Introduction

Persistent work on drying theory has brought little prospect that it will ever be able to predict dryer designs, and thus open a shortcut to practical solutions. When conditions are complex, theories cannot be made to predict reality. The theory of gravity, for example, predicts poorly the motion of as few as three bodies in space and gets worse as the number of bodies increases. In drying, too, the motion of fluids in solids is complex, and predicting is made more difficult because there are so many solids, each with different properties.

For these and other reasons there is little transfer of technology from drying research to practice. Tests are almost always needed to indicate how a material will behave in an actual unit. This is confirmed by various publications (Keey, 1978; Marshall, 1954; Mujumdar, 1987; Perry and Green, 1984; Reay, 1979). We therefore run tests to find out what works, and we use other practical methods, based largely on know-how and fundamental relations, together with a few basic material properties, to complete our calculations and designs.

1.1 Scope and Definitions

Of the roughly four million known substances, about 60,000 are processed and sold; many of these must be dried. Sales of chemicals alone have reached about $240 billion a year in the United States, according to the Chemical Manufacturers' Association. Nearly every industry has wet solids that must be dried, some more than once. Most of these are particulates—powders, grains, granules, crystals, pellets, flakes, chips, and other small forms. They range in size from a few microns (micrometers) to about one-half inch (1 cm), and they are generally of rough shapes; sometimes they are cylindrical or spherical. Their size and shape may be formed before the drying operation or by it. For example, when liquid feeds are sprayed, they become spherical droplets, which may expand as they dry and partially break up while drying or being conveyed and collected.

The great diversity of materials to be dried and the high cost of drying have resulted in the evolution of many dryer designs. Most are for particulates, and most operate continuously at about atmospheric pressure. For a better understanding, they can be divided by heating method into two types, as described in Dittman (1977). The definitions given here are for ideal conditions. Many actual systems are combinations of both types.

1. *Indirect dryers* use a hot surface to heat the solids by conduction or radiation. They use little or no hot gas such as air. This is a solid-liquid-vapor system, and the process is nonadiabatic, that is, heat is added from outside the system.

2. *Direct dryers* heat the solids by contacting them directly with a hot gas (usually air). This is a solid-liquid-vapor-gas system, and the process is adiabatic, that is, all the heat is in the system; none is added to or taken from it, assuming perfect insulation.

The method of heating determines how the drying solids will be agitated, transported, and collected. The split between heating methods is not sharp, however, because most indirect dryers use hot air sometimes, and some direct dryers use indirect heating. The overlap blurs definitions, and comparisons are difficult because few applications fit both types.

Another important distinction is the feed material's physical state. Liquids vary from thin solutions to thixotropic pastes, and solids vary from nearly dry granules to sticky sludges. Spray and drum dryers can be fed only liquids. All other common dryers accept wet solids, although certain liquids can be pumped into some of them, most often through a spray nozzle.

Except for the general discussion of internal and external drying conditions presented in this chapter, no effort is made to cover drying theory. Theoretical drying studies can be found in several texts, such as Keey (1978), Perry and Green (1984), Mujumdar (1987), and elsewhere. But available texts include few of the practical aspects that are dealt with here.

The drying operation is complex and involves a number of chemical engineering operations, including the following:

Vaporization	Diffusion
Flow of heat	Conveying
Flow of fluids	Filtration
Transport of fluids	Mixing

Psychrometry   Crystallization

Combustion   Fluidizing

Size separation

Also important in drying systems are the solids-handling operations—transporting, mixing, classifying, preforming, grinding, compacting, and packaging. The first three are often performed inside the dryer.

1.2 Why Drying Is Needed

Many materials are processed in the liquid state—ideal for mixing and reacting—but most products are needed or wanted as dry, or relatively dry, solids. Several operations can reduce moisture content at less cost than drying. Some liquids (usually solutions) can be concentrated by evaporating; insolubles can be converted to wet solids by decanting, filtering, or centrifuging. These operations often precede drying to reduce the moisture content, but seldom can anything but drying remove all or most of the moisture.

Classes of products dried in the process industries include polymers, foods, pharmaceuticals, minerals, agricultural products, wastes, ceramics, clays, catalysts, colorants, and various other organic and inorganic chemicals. The most common reasons for drying and some specific examples follow.

1. Preserving. Many solids spoil quickly in water, but last a year or more when packaged dry.

2. Reducing weight for shipping. Clays are dried for shipping, then redispersed in water for paper making.

3. Reducing weight or volume for packaging requirements. Many foods and detergents are dried to suit consumers.

4. Making specific shapes or uniform mixtures for further processing. Ceramics, mixed with additives, are dried into spheres of uniform size and composition for pressing.

5. Recovering solvents for reuse while drying. Abrasives are dried from methanol slurries.

6. Separating a noxious or toxic liquid from a solid.

7. Removing an unwanted solid and recovering the liquid.

1.3 Theoretical Drying Studies versus Design by Testing

For years fundamental research has been conducted on drying to improve understanding and to derive mathematical design methods. The main interest of this research concerns the internal drying conditions in order to learn what takes place inside solids as they dry. Although much information has been gained, even the theorists conclude that the equations thus far developed are not adequate, and tests must be run to get dryer design data (Dittman, 1977; Mujumdar, 1987; Perry and Green, 1984; Reay, 1979). Even further from reach is the more basic determination of which type of dryer, if any, is the most suitable for a specific material. This, too, requires testing.

Without tests there is no way to tell with confidence whether even closely related feeds will dry in the same manner and at the same conditions. This also applies to broad classes of materials (such as families of inorganic salts, polymers, and ceramics) that have similar chemical and physical properties. Testing is an empirical trial-and-error process that requires know-how. Test results are useful only for a specific application and as a limited guide for other tests, but there is no other way to get the needed information.

1.4   External Conditions

Those conditions that influence the drying but are outside the solid are termed external conditions. They concern the bulk flows of materials—solid, liquid, vapor, and gas (if any)—in contrast to the conditions of the microscopic flows that relate to internal activity. Some conditions concern both internal and external activities.

The important external conditions can be divided into two general groups for a given application.

1. Conditions usually fixed
   a. Properties of the materials
   b. Method of feeding, heating, supporting, and mixing the solids; also the means for transporting them throughout the drying system and removing them
   c. Characteristics of the drying vessel, its heating method, and its provision for removing the vapor
   d. Materials of construction and type of insulation
   e. Operating pressure (usually atmospheric)
2. Variable or elective conditions
   a. Heating temperatures and temperatures of feed, product, and gas
   b. Moisture content in feed, product, and gas
   c. Flow rates of feed, product, and gas, including any recycling of solids or gas
   d. Exposure time and distribution of exposure time between particles
   e. Feed pretreating or backmixing

f. Product transport, collecting, and conditioning (cooling, compacting, and grinding, for example)

Tests are run to find the best set of results. This is done by varying some of the external conditions. The most basic inputs are heating temperatures, exposure time, the flow rate of the solids, and their concentration in the feed. To get the desired results, the test unit may have to be modified, or the feed altered, or another type of dryer tested. The aim is to get design data for scaleup to a commercial system that meets all the product and rate specifications.

The most important of the conditions is the motion of heat. It is very diverse and is both an external and an internal condition. It is transmitted in three modes (McAdams, 1954)—in most dryers by all three, but usually one dominates.

1. *Conduction* moves heat between molecules from higher to lower levels of energy (heat) in solids (and to some extent in fluids). It is the principal heating mode for indirect drying, moving heat from hot metal to the particles and then between particles.

2. *Convection* moves heat between adjacent volumes in flowing fluids, inside of porous solids, and outside of all solids. It is the principal heating mode for direct drying.

3. *Radiation* beams heat through space from each object to all cooler ones in its line of sight. It is important—perhaps dominant, especially at high temperatures—in rotary dryers, but in most others its influence is minor.

1.5 Internal Conditions

Heat enters a wet solid being dried and evaporates some or all of the liquid, thus forming vapor, which is driven out. Heat transfer and mass transfer, and the forces that drive them inside the solids, together with the properties of all the components, are the internal conditions.

Heat and mass transfer occur simultaneously as solids dry. The mechanisms of mass transfer in solids are listed here. At times one will be more active than the others, and, depending on conditions, they may oppose or help each other.

1. Converting liquid to vapor expands the volume (about 1500 times for water in a typical dryer), thus expelling the vapor.

2. Capillarity is the flow of liquid through small pores and other small openings within and around particles. It is induced by molecular forces strong enough to overcome gravity.

3. Diffusion in drying is the flow of liquid or vapor within porous

particles, induced by changes of temperature and pressure. It is also the moving of vapor out of a porous solid and mixing with a gas.

4. Pressure differentials in the fluids in the pores of a solid impel motion. They are caused by temperature differences and by shrinkage and other physical changes during drying.

5. Gravity plays some part in all these actions, most notably in static solids, as in conveyor or tray dryers.

6. Evaporation and condensation are ongoing processes in which condensation in cooler areas and reevaporation move heat and create pressure differences that induce fluid motion.

The transfer of heat into the solid and its entrained liquid and the resulting mass transfer are a cause and effect relationship. The evaporation and other actions that occur are affected by the presence or absence of gas.

For indirect drying little or no gas is present, and the solid and its liquid are heated above the liquid's boiling point. Evaporation causes some degree of expanding, contracting, shrinking, and cracking of the solids, in addition to the great expansion of liquid to vapor. There is also some recondensing of vapor on cool particles—only a few are in contact with the hot surface. In this operation the primary driving force is a temperature difference, and the result is mass transfer. Liquid and vapor move inside the solids by capillarity, diffusion, pressure differences, and gravity. Vapor is moved into the space outside of the solids by pressure differences and diffusion.

When a gas is present, the solid and its entrained liquid are also heated, but to a temperature below the liquid's boiling point. But the lower-temperature driving force is more than offset by the vapor-pressure driving force. The higher vapor pressure of the liquid drives vapor into the gas, which has a lower vapor pressure.

Drying stops when the two pressures are equal. For short exposure time drying this equilibrium is approached but not reached. For any drying, if there is bound moisture in the solids, its vapor pressure is lower than for free liquid, and the equilibrium partial pressure of vapor in the gas is correspondingly lower.

When a porous particle dries, liquid at its surface evaporates, and the amount of remaining liquid as well as the ease of its release depend on the nature of the solid. The full drying range has three overlapping stages in which heat and mass transfer are resisted to varying degrees.

First, liquid vaporizes freely from the exposed wet surfaces at a constant and relatively high rate. Then the rate slows as capillary action within and between the particles brings water to a partially dry surface. Finally, it slows further as the solid warms, and vapor that forms below a completely dry surface must diffuse out as heat flows in. The rate of drying falls in each of the last two stages. The limiting resistance changes from heat transfer in the constant-rate stage to mass transfer in the falling-rate stages.

Indirect drying.   Indirect drying is simpler than direct drying in concept and with regard to equipment systems. It is widely practiced but receives far less research or published attention. Its basic operation is to conduct heat into particles by the hot metal walls of the containing vessel and agitator, if any. The particles that are heated in turn heat others—slowly because heat conduction is low in most solids and is retarded by porosity. Radiation plays a minor part. Probably of more importance is the condensation of some vapor on the outside of or within cooler particles.

Drying is faster and more uniform when agitation brings cooler, wet particles continuously into contact with the heating surface. To provide a temperature driving force, the temperature of the heating surface must be above that of the solid, which, in turn, must be above the liquid's boiling point to assure evaporation. To avoid heat damage to the product, it is often necessary to resort to the use of some air or to vacuum operation.

Direct drying.   Direct dryers, on the other hand, heat the solids by a hot gas, nearly always air, which passes over or through the solids or supports them. Surrounded by hot air, the outside of each particle is heated more uniformly than by indirect dryers. Heating is further improved if the solids are porous, because the hot gas enters as the vapor, which is nearly always water vapor, leaves.

When the flow of solids and air is cocurrent (the most common mode), the solid and its entrained water normally remain well below the water's boiling point. This is because the pressures of the air and water vapor combine to form the total pressure, with the vapor contributing the smaller part. Thus the water evaporates at a low pressure and temperature, even if the air is quite hot. This keeps the solids relatively cool as long as they retain some moisture. For most applications the product is discharged at some temperature level between the evaporation-cooled temperature and that of the outgoing air.

1.6   Feed Properties

For nearly all drying applications the greatest supply of heat is required for the liquid's latent heat of vaporization (except for feeds at very low moisture contents). Other influences are the liquid and vapor specific heats, the solid specific heat, and, occasionally, the heat of crystallization. The latter is equal in value but opposite in sign to the heat of solution. It can either add to or subtract significantly from the material's heat load, but it is negligible for many substances, and so it is often overlooked.

Dryer feeds are either wet solids or solids in liquids (solutions, slurries, sludges, emulsions, thixotropic pastes, and others). The main concern, however, is wet solids because, in drying, even a heated spray or a thin liquid coating quickly turns solid.

Most solids are porous (even polymers, for example, are usually porous until after fabrication), and liquid trapped inside flows through capillaries to reach the surface and vaporize. While drying, some materials lose porosity (as when the rising temperature causes some fusing), slowing the further release of moisture.

To some extent many solids are hygroscopic, that is, liquid is held tightly enough, chemically or physically, to keep its vapor pressure below that of pure liquid at the same temperature. Some or all of this bound moisture, together with unbound moisture, can be removed by drying. The extent depends on external drying conditions, such as time and temperature. The total moisture removed under specific drying conditions is called free moisture.

Some wet solids are thixotropic, that is, they become fluid enough when worked, as by pumping or other shear forces, that they flow and can be sprayed. Others are dilatant; they become stiffer when subjected to those shear forces and are thus more difficult to handle. Many solids are sticky and require backmixing with dry solids to give them a consistency suitable for feeding.

Ordinarily dryers cannot accept feeds unless they flow freely. A liquid needs the viscosity, surface tension, and discrete size of particles (if undissolved) that allow it to be pumped and, if necessary, sprayed. A wet solid usually needs the consistency, dryness, and particle size and shape that allow it to flow properly (generally measured as angle of repose); otherwise it has to be mixed with enough dry material to make it so.

Additional properties that affect drying include the following.

1. Cohesion and adhesion govern "stickiness", namely, how a material sticks to itself and to equipment surfaces.

2. A few materials pass through a tacky, or sticky, phase as they dry or become plastic with high viscosity.

3. Particle size affects the speed and uniformity of drying. Thus it influences the type and size of dryer and product quality.

4. Some materials skin over or case harden as they begin to dry,

and the resulting film may seal in the remaining moisture, at least partially or temporarily.

5. Heat of crystallization can increase or decrease by 20 percent or more the heat required per unit of evaporation and cause a similar change in the airflow rate.

6. Impurities, even in trace amounts, can seriously affect the performance of a dryer. They may require changing to another design or, more likely, altering the feed, such as changing the pH.

1.7   Effect of Drying on Properties of Solids

Wet solids and solids in liquids vary greatly in chemical, biological, and physical properties, and when dried they sometimes undergo surprising changes. The following are the main influences on final products from wet solid feeds.

Properties of the feed liquid and solid, including the components and structure of the solid, especially its sensitivity to heat and moisture

Heating method

Heating temperature and length of exposure to heat

Initial and final moisture contents

Type and degree of agitation or turbulence

The various operating conditions also govern both product acceptance (meeting specifications) and efficiency (energy use and productivity). Unfortunately the most profitable operation is often on the verge of trouble because the best fuel economy and production rate are obtained at the upper temperature limit, just below the point of fouling the dryer surfaces, scorching the product, or creating the circumstances that allow a fire to start.

Backmixing dry powder into a bed of drying particles, or spraying water into the bed, sometimes agglomerates particles. Conversely, the particle size is reduced when fragile materials break up, even in the gentler indirect dryers, but far more in those direct dryers that air-convey the dry product, in particular through sharp bends in equipment and ducts.

There is an even greater transformation when sprayed liquids are dried, and their final products, particularly from spray dryers, are influenced by the following additional factors.

Atomizing method and type of drying chamber

Air moisture, especially at the product outlet

Inlet air temperature (affects degradation)

Outlet air temperature (affects product moisture and other properties)

Product collection method

A number of product properties, and their uniformity, may be vital to a product's ultimate use and thus to its acceptance. Some of the more important ones follow.

Absence of degradation (such as scorching or other chemical, physical, or biological change)

Moisture content within a specified range

Particle size and distribution

Minimal dustiness

Bulk density

Flowability

Degradation is most affected by the temperature-time relationship in the dryer. In an indirect dryer a product's heat sensitivity sets the upper temperature limit. The exposure time may also be limited, or to some extent tied to the temperature. But in a direct dryer the limits are not as easily identified. In the usual case of cocurrent flow, the temperature of air in contact with wet particles can be far above their heat sensitivity limit. Evaporative cooling protects the product, but some particles—fines in some dryers, coarse particles in others—may stray onto hot metal and be burned.

Other causes of degradation include any tendency of wet solids to case harden or of liquid droplets to form films. These conditions often are unavoidable, and they are most likely to occur with fast initial evaporation caused by high temperature. They restrict continued evaporation, and this may affect the final product moisture or uniformity.

Lower moisture in the product can be reached by higher temperature at the outlet, by longer exposure time, and, if air is used, by lower air humidity. Lower moisture content than specified, however, is seldom desirable. It raises energy use and lowers productivity. Moisture can agglomerate crystals or bind the components within particles, inhibiting such problems as attrition and dustiness. For some products moisture adds salable weight.

In contrast, an excessively high moisture content may fuse particles into a block, making the product less acceptable at the very least.

Sometimes blocking is delayed, as when a product is dry on the surface when packaged, but over time excessive internal moisture migrates to the surface.

For wet solid feeds, the particle size depends mostly on the original size in the feed. But many materials shrink on drying. This can slow drying by closing pores, and it can prevent meeting size specifications. Also, some particles break up either from the drying action or while being conveyed. For both solid and liquid feeds, some properties are affected by others, such as bulk density and flowability by particle shape, size, and size range.

For a liquid feed, whether slurry or solution, the particle size depends on droplet size, size range, and what the feed is sprayed into. A coarse spray into a bed of semidry particles enlarges them. But atomizing into hot air in a chamber normally limits the particles to an average of no more than 150 microns (μm); they may be spheres or irregular shapes. If spheres (the more common), they are nearly always hollow, may expand or contract, and then may break. Expansion results when droplets film over, evaporation is delayed, and the captured vapor (or gases escaping from the liquid) are heated (Marshall, 1954). Liquid released from spheres that are exploding or imploding as they dry, adds to product fines.

The size also depends on the type and degree of atomization and on the properties of the feed (in particular, the viscosity, surface tension, and concentration of dissolved or suspended solids). One illustration of the diversity resulting from atomized liquids is the tendency of some to form one or more satellite droplets off the main drop, which are occasionally captured by the main droplet and fused to it.

Bulk density determines whether or not the prescribed weight of material will fit into its designated container. Because small particles fill the voids between large ones, bulk density is affected by size and size range. A higher proportion of small sizes increases density. Also, flat shapes are generally denser than round or irregular ones, again because of fewer voids. The low density of hollow spheres is increased if some break in the process.

Many products must be free-flowing, a property that depends on particle shape, size, size range, and freedom from any stickiness. Larger sizes are less dusty and dissolve faster. Dry spheres of uniform size flow readily and distribute evenly, but good control over particle size, flow, or density cannot be obtained in dryers, except to a limited extent in spray dryers.

There are various other properties, of physical, chemical or biological nature, any one of which may be the most important for a particular product—color, flavor, dispersibility in liquid, aroma, or others. Most applications have several property specifications. Sometimes the external conditions have to be arranged to compromise conflicting requirements.

1.8   External Conditions and Dryer Design

To complete an acceptable dryer design, much engineering and commercial information is needed, plus the right external conditions; some are fixed by the material or the type of dryer, others are found by testing. Exposure times vary from less than 1 s to well over 2 h. The use of air extends from none at all to supplying all heat and support of the solids. The motion of solids in dryers ranges from a motionless ride on a tray or belt, to mechanical agitation, to turbulent conveying in air.

Every dryer has some design limitations. All have a practical upper limit of size (and cost); some have a design-limited exposure time; all have temperature restrictions. Some indirect dryers have to be designed to use airflow or vacuum to keep the evaporation temperature down. Direct dryers, on the other hand, use large amounts of air and thus need more complex systems. The many ways that air is used in drying systems are a central part of the chapters that follow.

2

Drying Systems

2.1 Introduction

At a New Jersey dyestuff plant—when labor practices were lenient—an operator gave a remarkable daily performance running three spray dryers. Each was spread out on three levels, and he had to know the status of all, though they were at different stages of operation and were drying different products.

To start such a dryer (which is illustrated in Fig. 2.2), the operator opened the air heater's steam valve on the second level and then started the feed pump and atomizer on the third level. Back on the second level, he slowly opened the feed valve, waiting for the right inlet and outlet air temperatures. In between times he checked the other units, and he had to know when to exchange each of the three product drums on the first level.

A typical run was only 1 to 2 h. Then he completely brushed down inside the chamber and duct work and kept the brushings separate for later reworking. Colors for each series ran from lightest to darkest, after which he thoroughly hosed down the chamber and ducts, about a 1-h job. In spite of a hectic schedule and all the stair climbing, he never seemed to be rushing.

2.2 Costs

That system of three dryers operated successfully for many years and underwent only two important changes. The drying chamber outlets were modernized, and product collection was expanded to meet stricter pollution rules. The old standby designs have not, in general, been improved much over time, nor have new designs replaced them in any significant way. As a result of rising costs, however, there has been some shift toward dryers that need less energy or maintenance.

There have also been some changes in working environments. As an example, operators spend less time monitoring dryer conditions on systems with computerized controls.

In the early stages of a project, only safety is of more concern than whether a dryer can make the desired product. But in later stages cost dominates. This includes the cost of equipment and installation and the cost of operation, mainly energy. For buyers, systems must operate efficiently and avoid expensive downtime. For suppliers, equipment must be designed cost-effectively and accurately because, from their standpoint, success in the various stages—engineering, fabricating, erecting, installing, and start-up—depends on the test and design efforts.

The drying costs per unit of product can seldom be compared on an equal basis, but they are generally higher for indirect dryers. Their limits on drying temperatures restrict capacities and raise heat costs. In addition, they nearly always use indirect heating, while the more efficient direct-fired heating is often used on direct dryers. Labor expenses for operating are about the same per drying system, but for indirect dryers they are more per unit of product. In addition, total labor time is greater because of the mechanical drives and tolerances, especially for drum dryers.

When designing a dryer, the usual worth assigned to it when its original use ends is mere scrap value, even though its fate may be different when that application is over. It may be offered to the used equipment market or cannibalized for parts. But higher profits could be realized if the system were applied to some other use, and it costs less to plan for such adaptations before rather than after the dryer has been built.

Another drying application would, in most cases, bring the best return. But drying equipment can also perform other operations, such as cooling, congealing, reacting, heating, sterilizing, agglomerating, encapsulating, crystallizing, concentrating, coating, and calcining. Fluid-bed, rotary, and indirect units specialize in heating and cooling. Rotary and flash dryers are commonly used for calcining and other high-temperature processing. While these nondrying uses normally have lower paybacks, some could be attractive, especially if designed for it from the start.

2.3 Design

A typical drying project takes 6 to 12 months from concept to start-up. It begins with testing, followed by the design stage after a contract is signed. Sometimes further tests are needed. The external conditions determined by testing, together with engineering data and know-how, form the basis for designing a system and selecting its components.

The lowest cost for dryer and components is generally found at the smallest volume, consistent with fulfilling the design within safe limits. Some safety factors once designed into drying systems have been eliminated by competitive pressures. But if bottlenecks are to be avoided, there must be some overdesign in such areas as temperature limits, heat loads, fan capacities, or whatever is critical to the current job and to any known future operation.

Air use—problems and benefits.   A major factor in equipment design, operation, and cost is the amount of air (or other gas) used for heating. It is best kept to a minimum. For some indirect drying jobs it is eliminated except for unavoidable leakage. For direct dryers it supplies the heat, and in some it suspends the solids while drying them.

Indirect dryers often need some air to sweep out the vapor. Usually the air enters at the product discharge and leaves at the feed inlet. If the solids cannot withstand a temperature above the boiling point, enough air is needed to lower the evaporation temperature. The solids are kept relatively cool by evaporation, most effectively when air and solids flow in parallel. Counterflow, on the other hand, allows reaching higher solids temperatures, thus obtaining lower moisture levels.

Three major costs offset the benefits of using air—the extra air-handling and product-collecting equipment, the heat lost in the exhaust, and, when necessary, treating the exhaust to keep noxious gases or dusts out of the atmosphere. There are other disadvantages. Large volumes of air can carry off desired components, such as taste or aroma. Combustion gases damage some products, as can oxygen, which supports conditions for fires or dust explosions in vulnerable parts of a system. To an extent these situations can be improved or corrected by proper designs, but costs can be high.

Temperature and exposure time.   The basic operating conditions are temperature and exposure time, which are usually interrelated. One or both are used to control the drying, but only within certain limits. Upper limits on temperature are set by the product's heat tolerance over time, and by the metal's cost and loss of strength at higher temperatures. Exposure time is also limited by the product's heat tolerance and by equipment design. For example, the time of the airflow through most flash dryers is less than 3 s, 20 s for spray dryers.

Because indirect dryers heat the drying solids above the liquid's boiling point, the metal contacting the solids has to be above this temperature. The risk of heat damage depends on contact time and temperature. For direct dryers, on the other hand, the solids are affected by time and by air temperatures, mainly at inlet and outlet. In most direct dryers, however, the solids are not heated much above the wet-bulb temperature, which is well below the boiling point.

2.4   System Configurations

Indirect dryers.   Indirect dryers have a relatively simple flow pattern, as illustrated in Fig. 2.1. The diagram shows a disc or paddle-type dryer, with the vessel sloped to aid transport of the solids. Heat sources to jacket and agitator could be separate flows, with the jacket in some cases zoned for different temperatures, sometimes cooling in the last zone.

In this system, leakage into the vessel is minimized by the use of choke-type feed and discharge screw conveyors. These cylindrical units run full and block off the entry of air. The little air that does leak in is drawn out with the vapor, and in most cases a condenser is not needed.

Direct dryers.   For most direct dryers heated airflow is in the once-through, or open, mode. This air is heated and drawn or forced, or both, through the vessel. The dried product is separated from it in one or more collectors. The air is then either discharged to the atmosphere or partially recycled. Fully clôsed systems that recycle all the gas—usually nitrogen—are described in Chap. 13.

Once-through systems.   Figure 2.2 illustrates a simple once-through process, and the spray dryer shown could be any direct dryer. Most use the same basic layout with air heater, feeder, drying chamber, product collectors, blower, and connecting ducts. Some designs filter the inlet air; some have only one product collector. The cyclone-scrubber arrangement shown is common, but the wet fines—in the form of a sludge—must be reworked or disposed of.

Most direct dryers have automatic control of the process, which includes safety sequencing of heater, feeder, blower, and other elements. To overcome relatively high pressure drops, fluid-bed dryers and some

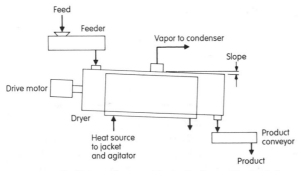

Figure 2.1 Indirect dryer with in-leakage blocked by feeder and product conveyors.

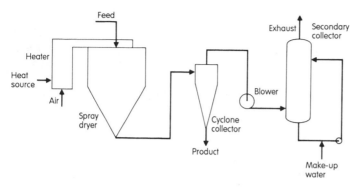

Figure 2.2 Once-through direct drying system.

others often have two blowers in what is called a push-pull arrangement.

Figure 2.3 is a more complex once-through plan that is one of many possible layouts. Flash drying in particular needs this kind of versatility. On many applications products are size-separated by one of a variety of classifiers into coarse and fine fractions. The coarse is returned to the dryer to give it more time to dry. In this arrangement all of the coarse fraction and some of the fines are backmixed into the wet feed to improve its consistency. The final product is the remainder of the fines from the primary collector mixed with the powder from the secondary collector.

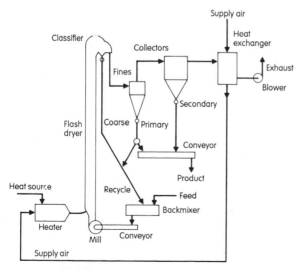

Figure 2.3 Direct drying system with backmixing, milling, classifying, product recycle, and heat recovery.

The layout in Fig. 2.3 has certain optional items, and more could be added. Following the dual collectors is a heat exchanger to preheat the supply air with the exhaust. Many installations for foods or light-colored products need an inlet air filter to keep out dirt and specks. Systems can be tailored to suit the requirements for preparing the feed, conditioning the product, and treating the exhaust or extracting heat from it. Other possible variations include a solids conveying loop for product cooling, heating, or secondary drying.

Recycle systems.   Figure 2.4 shows a scheme that recycles most of the air, thus reducing the amount of exhaust. In addition, the direct-fired heater converts oxygen to carbon dioxide and water vapor. The process is called self-inertizing, because the fraction of inert gas is increased as the oxygen content drops, reaching that of the gas out of the heater. For some duties it can be reduced from the air's normal 21 percent by volume to less than 10 percent. This helps to preserve products that would be oxidized at higher levels. A calculation method for this oxygen content is given in Sec. 5.2.8.

All materials entering and leaving the system are kept in balance. In the layout in Fig. 2.4 the weight of dry combustion gases and excess air entering from the heater, plus any in-leakage and bag filter purge, equals that of the dry gas bled off. The total water entering is the evaporation plus water vapor from the heater, and that amount is removed by the bleed and the scrubber condenser.

Less exhaust costs less to treat, should that be required. Incineration is a common method, often mandated for pollution control to eliminate noxious powder or gas. But fuel must be burned to heat the conveying gas to the vapor's ignition temperature, and additional air may be needed. Exchanging heat from the incinerator's exhaust to its inlet

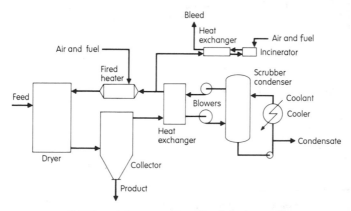

Figure 2.4 Self-inertizing, recycling direct drying system.

stream can lower its fuel demand. But its burner needs enough turn-down to avoid heating the inlet stream past its ignition point while still in the exchanger. Systems with explosive mixtures must be de-signed to maintain low oxygen levels and avoid air in-leakage.

The exhaust-to-supply air heat exchanger, also shown in Fig. 2.4, is a worthwhile option. An efficient driving force is provided by the cool gas from the scrubber on one side of the exchanger, taking heat from the dryer exhaust on its other side. It requires a collection system, however, that prevents fines from passing through, which might foul the exchanger surfaces. Under some conditions moisture can condense out of the exhaust and cause or worsen fouling or corrosion, but this can be avoided with proper temperature control. Without the ex-changer, the total heating and cooling costs for recycling in this way are usually somewhat higher than for a once-through system.

2.5 Major Components

2.5.1 Feeders

Spray and drum dryers accept only liquids, but viscosities can vary over a wide range. Most other dryers need somewhat granular feeds, although some pastes, or even slurries (when backmixed with enough dry product, or otherwise pretreated) can be brought to the right con-sistency. Wet feeds can be backmixed directly in a few direct dryers if they provide a bed of acceptable dry material. But most wet feeds must be kept from dropping immediately onto a hot surface.

For wet solids a number of feeders are suitable, and variable drives can be used for control. The mixer-type feeder serves as both feeder and external backmixer. The agitation in some dryers gives a degree of backmixing without adding dry powder. When this is possible, it eliminates the need for a bigger dryer.

The types of feeder commonly used for wet solids, in addition to mix-ers, are screw conveyors, rotating tables, vibratory trays, and rotary air locks. Some of these have a hold-up characteristic that provides a relatively uniform rate. Another way to even out the flow is to install one in tandem with the screw type. Dryers with wide beds need special feeders that spread the material evenly to assure uniform drying, as on conveyor dryers and plug flow and vibrating fluid-bed dryers.

2.5.2 Heaters

Heaters, as well as dryers, can be classified as either direct or indirect. Direct refers to heating air by mixing it with combustion gases, while indirect is heating the air (heating the solid in the case of drying) through the metal wall of an exchanger.

The cost of direct-fired heating is less, but some products are damaged by combustion gases. On the other hand, direct firing helps preserve those products that need a lower oxygen content, which is especially low in systems that recycle some of the air. Heating by methods other than direct firing cannot escape some inefficiency. They require either a heat exchanger at the site or a remote boiler, as for generating steam or electricity.

The range of heater types in common use for dryers and their usual temperature limits are shown in Table 2.1. A combination of heater types is often cost-effective, such as a steam preheater before an electric heater. It is also possible to reduce costs or avoid shutdowns by providing standby oil service for a gas-heating installation, for which combination gas and oil burners are used. Indirect heaters have low thermal efficiencies, which can, however, be greatly improved by exhaust heat recovery exchangers.

Most products have a temperature threshold above which degradation begins. Some can withstand very high heat, however, and their drying temperatures are limited only by the temperature-strength relationship of the heater walls. Some heaters, such as those for calcining, are lined with refractory to allow maximum heat. The upper limit of air temperatures for construction using the common stainless steels is 1200°F (649°C); it is higher for certain alloys. Intermittent operation is more severe than continuous operation.

Heating, more than any other factor, makes drying expensive, and wasteful practices increase the cost further. Losses and cost factors vary with the type of heater and with the fuel and method of converting and delivering it to the dryer. Transmission or transportation costs are high, on a decreasing scale, for electric heat, steam, and thermal oils.

Most indirect and some direct dryers heat with steam, which bears a penalty for heat lost in returning or discarding the condensate. Direct dryers that use direct-fired heating avoid those inefficiencies. But for all direct dryers the exhaust air carries out much of the heat. Roughly 35 to 55 percent of that loss can be recovered using an exhaust-to-supply air heat exchanger.

TABLE 2.1 Common Heaters for Drying Systems

Type of heater	Heat source	Maximum air temperature	Remarks
Direct	Fired gas, oil, etc.	1200°F	Adds H_2O, CO_2, etc.
	Electricity	600°F	Convenient, high cost
Indirect	Fired gas, oil, etc.	800°F	Stack heat loss
	Steam	400°F	Convenient
	Thermal fluids	800°F	One or two phases

Heat flow from an indirect heater is limited by the three factors in the fundamental equation

$$Q = UAT_m \qquad (2.1)$$

which shows the heat flow Q to be a function of the heat transfer rate U, the surface area A, and the mean temperature difference T_m. From this equation it can be shown that a steam heater of practical size can heat air only to within about 20 to 25°F (11 to 14°C) of the saturated steam temperature. (Superheated steam contains too little heat to be useful.) Other indirect heaters are similarly limited.

Electricity is the most convenient heating method, but cost rules it out of most moderate to high production rate jobs. Suitably controlled electric heaters have the advantage of unlimited turndown, but in steps. A unique, but potentially troublesome feature of electric heat is its constant energy input. As Eq. (2.1) shows, at constant heat flow and area, a drop in the heat transfer rate causes the temperature to rise, which in turn can cause fouled electric elements to overheat and fail.

Hot waste gases from other processes can sometimes be used as heating media for dryers. But moisture contents are often so high that the desired product moisture cannot be reached. The high humidity can also cause condensation on cooler parts of the system.

2.5.3 Product collectors

An unavoidable and costly operation is separating product from the airstream, usually the final step in a drying system. It can also be an intermediate step, as in the use of a cyclone to prevent the escape of fines from a fluid-bed dryer. When fines fractions are collected separately, they may be added to the main product, reworked, or discarded. Some collectors and the ducts leading to them also provide extra drying time, which is significant for the short exposure time of spray and flash dryers.

There are several ways to separate powder from an airstream, and many equipment types. Scrubbers alone are too numerous and diverse for a rational classification. Four basic types are common in drying systems—the drying vessel itself, cyclones, bag filters, and wet scrubbers. There is some competition between types, but in general each has its own niche. Each operates on a different principle; namely, settling, centrifugal force, filtration, and, for scrubbers, a combination of actions. To assure full collection or provide separate fractions, two in series are often used, nearly always two different types.

Fires and explosions are infrequent in drying systems, but those that do occur are most likely to be in the collector. Thus explosion doors and other safety features are sometimes mandated. Powder buildup in collectors, or high powder concentration in the air inside them, can provide the circumstance for a fire or explosion that may be set by a spark or static electricity.

A more common problem is condensation, which occurs most often in systems that operate near saturation. Potential sites are those that are cool for lack of insulation. Condensation can be reduced with proper insulation, with gradual shutdowns, and, in the most vulnerable places, with spot heating devices.

Other important factors relating to collection are efficiency and pressure drop. In spite of many studies, predicting these is uncertain even for the simplest cyclones. Collection efficiency is to an extent a function of particle size, size range, and density. Also, powder loading in the airstream may help or hinder efficiency, depending on the collector type. Leaks, too, even small ones, can seriously reduce efficiency. For a given application the only way to find the best design and features is by testing, but samples should be representative of all the particles to be collected. Product from a cyclone, for instance, does not contain the fines that are the key portion for a test of a scrubber or bag filter.

Settling. Discharging the product directly out of the vessel is the obvious collection method for both direct and indirect bed-type dryers, because the product can readily drop out by gravity. A more active settling occurs in spray dryers for ceramics. In these and some other dryers, the airstream removes most of the fines, separating them from the coarse fraction. They are then taken out of the airstream in one or more collectors.

Cyclones. The centrifugal force in a cyclone throws the particles to the outer walls, where they slide down for discharging. Cyclones are inexpensive, but they subject the solids to abrasion, and the collection efficiency is low. A major advantage is the ease of clean-out with steam or pressurized water, using detergents when necessary. Also, they can withstand high temperatures and occupy relatively little space. In general, those designed for higher pressure drops have higher collection efficiencies, but air velocities must be within a limited range for reasonable efficiency without abrasion.

A bag filter or scrubber nearly always follows standard-sized cyclones because of their low efficiency, which declines as diameters get larger. High-capacity systems often use two or more smaller cyclones in parallel to boost efficiency. The multiple-cyclone design has a battery of tiny cyclones, 4 or 6 in (10 or 15 cm) in diameter. Because of its higher efficiency, it can usually operate without a backup. But uniform air distribution is needed to avoid reentraining powder. Air entry also affects standard cyclones if the transition into the spiral is not a continuous, smooth path.

The efficiency is reduced if air leaks in at the bottom of the cone, or if the true conical form becomes warped, rough, or dented. Leaks at the outlet can be minimized by designing with a bottom hopper and discharging through either a rotary air lock, a double-flap tipping valve, or a screw conveyor with spring-loaded choke.

Bag filter collectors.   The fabrics of these units are woven or nonwoven and are usually in the form of bags. The bags are cleaned in place by mechanical shaking, reverse air cleaning, or pulsing with either compressed air, compressed drying air, or nitrogen. These actions dislodge the particles primarily by shaking or flexing the cloth, the timing of which may be critical. Fabrics are chosen for temperature resistance as well as for filtration properties. Modular construction helps reduce unit costs, but makes selection more difficult for a specific capacity.

For many jobs the collection efficiency is high enough that bag filters can be the final, or only, collector. For the drying of colors and some hazardous and other materials, however, a scrubber or other backup is essential to prevent powder escape in the event of bag breakage. A cyclone placed before a bag filter may reduce bag abrasion, but the removal of coarse particles can adversely affect filtration.

Effective filtration depends on the coating of dried particles on the fabric. Thus the pressure drop through clean fabric starts relatively low and levels off, generally at 5 in WG (water gage; 1.2 kPa), when the particles added to the fabric equal those dislodged. Excessive pressure drop indicates that the fabric is blinded and needs to be cleaned, either by more frequent shaking or pulsing or by removal and wet or dry cleaning. Sonic vibration equipment can be used to enhance bag cleaning, and its gentle action can increase fabric life. The holdup of product in the fibers rules out bag collectors for some materials.

The usual design factor for drying is an air-to-cloth ratio of 4:1, that is, 4 ft^3/min per ft^2 of cloth area (1.2 m^3/min per m^2 of cloth area). Thus 250 ft^2 of fabric is needed for every 1000 ft^3/min of airflow. Backup units may have ratios as high as 10:1 ft^3/min per ft^2 (3:1 m^3/min per m^2). Ratios this high were once used for many primary collection duties, but for most powders they were found to be too high. Failure of bag clamps or of fabric is an occasional problem. A drop in pressure in a primary bag filter, or a pressure rise in a backup unit, indicates bag breakage in the primary unit.

Scrubbers.   Scrubbers are not collectors in the usual sense, because they yield a wet product, usually in the form of a sludge. Two of the many scrubber types are in common use in drying systems. Venturi designs have higher collection efficiency, but higher pressure drop. Impingement tray types are lower in cost and have many uses besides collecting fines, such as preevaporating feed, absorbing noxious gases, and condensing vapors, as in recycle systems.

The main disadvantages of scrubbers, in addition to not yielding a dry product, are a tendency with some products to foam or to cake up with solids, especially when concentrating the feed. Some foams can be controlled by additives, some by mechanical devices. The need to dispose of the effluent is eliminated if it can be redried. An occasional problem in cold areas is the icing of roads by mist from the exhaust.

2.6 Summary Data Sheet

Figure 2.5 summarizes the data for a drying system. While specifically intended for direct dryers, it can be used in its current form or modified to suit indirect dryers as well.

Figure 2.5   Drying system data sheet.

PROCESS AND EQUIPMENT					
CLIENT AND PERSONNEL					
Address, telephone, fax					
PRODUCT: Proprietary and chemical names					
Component	Feed, lb/h (kg/h)	%	Product, lb/h (kg/h)	%	Feed Consistency
Solid					Temperature, °F(°C)
Solvent					Density
Totals					Product size range
EQUIPMENT (No., type, materials of construction) Drying vessel					
Heater and heat source					
Feeder					
Collectors					
Other equipment					

OPERATION Airflow, ft³/min (m³/s)	Exposure time, s	Air temperatures, °F (°C) Inlet Outlet
Evaporation, lb/h (kg/h)	%RH (T_{dew}, %ASR)	Scrubber outlet, °F (°C)
Heat load, Btu/h (W)	Power, hp (W) Fan Feeder	Mill Total
Other data		

REMARKS	FOR COMPANY USE Date
	Project No.
	Personnel

3

Indirect Dryers

3.1 Introduction

Indirect dryers use little or no air (or other gas), and they heat the solids through a metal wall rather than by direct contact with hot air. Compared to direct dryers, they have lower maximum drying temperatures and lower maximum throughputs. In addition, they agitate the solids mechanically and have heavier construction. Usually operation is just below atmospheric pressure, as with direct dryers, but some are built for vacuum operation with pressures as low as 50 mmHg (7 kPa).

Dryers that do not use air must heat the solids above the boiling point of the liquid. But there is often a limit to the upper temperature set by the rigid construction. Thus temperatures above those reached by steam heating are generally not practical.

The most common indirect dryers have horizontal housings. These are either rotating cylinders or stationary housings with one or more rotating agitators. Housings for the agitator types are either cylindrical for optimum mixing, or trough-shaped to provide more area at the top for vapor removal. Major components are constructed of heavy plate. This gives them a rigidity that, together with the simplicity of their overall systems, makes them relatively easy to install and, when necessary, to relocate.

3.2 Design Considerations

Solids mixing. The design of an indirect dryer is influenced by basic differences in how the solids are handled. Most of the common types heat and agitate a bed of material, but there are two major exceptions. Drum dryers lay a film of liquid feed on the outside of a heated cylinder, and the high-speed paddle dryer throws a layer of feed particles on the inside of a heated cylinder.

Bed-type dryers are built for either plug flow or mixed flow. In ideal plug flow the particles hold their positions relative to one another, with a minimum of intermixing. A disadvantage is relatively poor heat transfer. A few products require plug flow so that each particle gets equal drying time. But most products benefit more from intensive mixing, which continuously exposes cooler particles to the heating surface. When this mixing is kept local, short-circuiting of incompletely dried material to the discharge is avoided.

For both mixed and plug flow, proper control of the heating and transport of solids is needed. To achieve this, the following important design factors have to be considered.

1. For heat transfer dominated drying provide high heating surface per unit of volume; for diffusion dominated drying provide long exposure time.

2. Improve the contact between solids and heating surface to increase the heat transfer rate.

3. Distribute the drying effect uniformly throughout the bed by better mixing between particles.

4. Separate the heating and moving of solids so they can be controlled independently.

5. Minimize uncovering of the heat transfer surface, caused either by shrinkage of the bed during drying or by the head of solids needed to push the solids through the dryer.

Air use. In most indirect dryer applications the use of air is held to a minimum, often to mere leakage. But many applications need a small sweep of unheated air over the bed to remove the vapor and prevent condensation. This is usually exhausted away from the outgoing solids to prevent reabsorbing of moisture. Sometimes, to achieve the desired product moisture, a small flow of heated air is passed through the bed near the discharge.

When the product is heat-sensitive, a lower evaporating temperature is mandated, and the flow of air must be larger. This may carry out fines and possibly require a collector. Removal of fines prevents their becoming overdried, but alters the particle size distribution of the product and may create a remixing or handling problem.

When using little or no air, the water (or other solvent) evaporates into nearly pure vapor. Thus to achieve evaporation, the solids must be heated at least to the liquid's boiling point (much higher if there is bound moisture). The alternative is to use more air or else to use the more expensive vacuum drying.

Except for vacuum units, some air leakage is unavoidable. Up to about 200 ft^3/min (6 m^3/min) in-leakage is considered normal for large units with flat gasketed covers and conventional bearing seals. Most feeders allow some air to seep in through the feed. When air is used to sweep out vapor, its velocity is generally between 10 and 20 ft/min (3 and 6 m/min). The dew point of this air depends on the dryer outlet temperature, but for most applications it is above 140°F (60°C) and below 194°F (90°C).

In vacuum operation, a practical level of absolute pressure is about 1 lb/in^2, or 52 mmHg (7 kPa). This lowers the boiling point to 102°F (39°C). At low pressures, gasket surfaces have to be minimized and manufacturing tolerances reduced for better fits and seals.

Heating and heat losses. Most bed-type dryers have heated agitators, the shafts of which may also be heated. If housing jackets are zoned, they allow a measure of temperature control. In some units combinations of operations can be performed, such as heating, drying, and cooling. The heating medium is most often steam, occasionally thermal fluids; electricity is seldom used because of cost. Combustion gases have very low heat transfer coefficients and thus are rarely used for indirect dryers, except for radiantly heated indirect rotaries.

An indirect dryer using little air needs much less heat than a direct dryer, if compared on the same basis (which is often not possible). But the heating medium of choice, steam, has efficiency losses at its point of generation and in its transport to the dryer.

Heat losses are unavoidable, and from a practical standpoint some cannot be recovered. Except for heat leaving in the product, the only major heat loss for indirect dryers is proportional to their use of air and the difference between the air's exhaust and inlet temperatures. Radiant-convection loss from the housing is normally low because of the use of insulation, which is also needed to protect personnel.

Heat transfer rate. Agitation increases the rate of heat transfer because particles move faster against the hot metal and mix better in the bed. But there are upper limits to the speed of agitation. Excessive speed can result in the following.

1. Fluidization, which insulates particles by a film of air, thus reducing the heat transfer rate, sometimes below the rate without agitation

2. Uncovering of some heating surface

3. Uneven drying, caused by short-circuiting particles to the outlet

4. Reduced exposure time to some extent, depending on the dryer design

When heat transfer dominates, the most effective design is one with a high heat transfer rate and a large heating surface. When diffusion dominates, a long exposure time is more effective, especially when the final moisture must be a few parts per million. Thus the design of commercial indirect dryers has to consider these two modes. Scaleup is on the basis of the heat transfer surface or rate on the one hand and the time required on the other.

Solids transport. Numerous feeder designs have been developed to meet a variety of problems. The main problem for dryers is feeding sticky or very wet solids. For these materials vibrating tray and rotating table feeders allow the feeding to be observed. At the point of feeding most air in-leakage can be choked off by using rotary valves or screw feeders. Special troughs are designed to feed across wide conveyor belts.

When wet or sticky solids first contact the heating surface, they may stick or cause other problems. They can be preconditioned by backmixing with dry product, usually in a mixer outside the dryer, sometimes inside the unit, depending on the agitator design and the nature of the feed. Either way the vessel must have enough volume to handle the larger flow. In addition, either more heat transfer surface or more time is needed because the wet particles are spread out in the dry recycled material, and they contact the heating surface less frequently.

Material is transported through the dryer by one of the following means.

1. Agitator design and speed

2. Slope of vessel

3. Clips or vanes attached to agitators at preset angles

4. A head of material built up by the feed being forced in

The head of solids is generally controlled by a weir, but some short-circuiting to the discharge cannot always be avoided.

Control. The usual criterion for the end of drying is the final product moisture. For indirect dryers this is gauged by the product temperature. When the exposure time is long, the controlled result lags behind the action taken. This may require temperature measurements at more than one point, with intermediate trim control to different zones of a heated jacket. In bed-type dryers, rotaries in particular, the loading and exposure times are critical to efficient operation. Measuring them, however, is seldom either convenient or accurate.

Exposure time is influenced by the head of material, the slope of the vessel, and, in some designs, by the agitator speed. Not only the average time is important, but also its distribution between particles. The latter influences product properties, and it is affected by the mixing action, short-circuiting, and air flowing through the solids that carries off fines. Most products that are discharged with an uneven distribution of moisture between particles equilibrate in time to a uniform moisture content.

Dryers in which the exposure time is coupled to the agitator speed give very uniform drying times and treatment to each particle. Changing the speed also affects the heat transfer rate, however, and this limits the ability to control the product conditions.

Zoned temperature control can be important for materials that go through changes in consistency. These changes occur in many materials and can be demonstrated by plotting the position in the dryer against the power required (Root, 1983). After heating to the boiling point, most of the liquid evaporates out of the wet feed. Then the solid goes through plastic and shearing phases before reaching the dry granular phases. The peak power required may be three times that at the beginning or end of the cycle. Thus power needs cannot be extrapolated from tests on either wet feed or dry product.

3.3   Specific Indirect Dryer Designs

The descriptions that follow include the functions, features, and operating methods of indirect dryers in common use in the process industries. More detailed descriptions of these and other indirect dryers can be found in manufacturers' literature.

3.3.1   Disc dryers

These dryers have hollow, full-circle discs, closely spaced on one or more shafts. This arrangement provides both agitation and a high heat transfer surface per unit of volume or plan area. Tip speed is typically 200 ft/min (1 m/s), but can be as low as 30 and as high as 240 ft/min (0.15 to 1.2 m/s). In the two basic designs the feed flows either parallel to one or more shafts or perpendicular to several shafts. Multiple shafts may be separated for gentle mixing or intermeshed for more vigorous mixing. The housing is also heated and is usually fitted with multiple zones for better temperature control.

Feeds should be free-flowing. If they stick to hot surfaces, they must be backmixed with dry product. The feed rate can create a head of material at the inlet, helping to force the solids through the dryer. But this action uncovers some heat transfer surface toward the outlet; the discs generally run 60 to 90 percent covered. Shrinkage aggravates this uneven flow, but it can be partially offset by a weir at the discharge. Fixed breaker bars are often used between discs to break lumps, prevent the mass from rotating, remove surface buildup, and improve mixing and heating. Plows can be added to the discs to improve mixing, lessen plug flow, and aid material transport. The use of air varies from none to enough to fluidize the bed.

Most applications need a relatively long exposure time; it can be from 10 min to 1 h. Long exposure gives more uniform drying, because then mixing tends to be local with less short-circuiting. To move through the dryer, the solids squeeze between the discs and the vessel wall in a thin layer. This gives a heat transfer rate from the jacket that is equal to—sometimes greater than—the rate for the discs. The overall heat transfer rate is about half that of the high-speed paddle dryer.

3.3.2 Disc dryers with segmented discs

Dryers with wedge-shaped, hollow segmented discs that are curved around a shaft are also called paddle dryers, even though both agitator design and action more closely resemble those of disc dryers. One design, shown in Fig. 3.1, uses two or four shafts that rotate oppositely in omega-shaped troughs. The semicircular discs are intermeshed to aid in mixing and self-cleaning. A single shaft in a cylindrical shell is used for batch operation or for vacuum or pressure designs. The hollow shaft and the trough's jacket are also heated.

Each disc plows through the bed of solids sharp end first. The wedge shape and intermeshing action scrape harder against the particles than do flat discs, which helps to keep the discs clean and increase the heat transfer rate. It also promotes local mixing, which gives uniform drying with a fairly narrow exposure time distribution. Average exposure can be from about 10 min to 2 h.

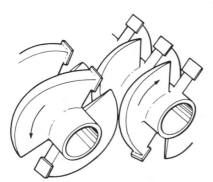

Figure 3.1 Segmented disc paddle dryer shafts.

Some very wet, even some sticky feeds can be handled because of the agitator action, but difficult feeds must be backmixed before drying. The mass is pushed through the dryer, not by agitation, but by clips on the broad ends of the discs.

Coverage of the agitators is ideally 90 to 95 percent, but may be dropped to as low as 50 percent by lowering the weir at the discharge end. This creates a head differential, which may be needed either to shorten the exposure time for a material with poor flow properties, or to adjust for the changing volume of the solids.

3.3.3 Drum dryers

Drum dryers may have single or, more commonly, double drums for producing flakes. This product may be too large to be considered particulates, but it is easily reduced in size. Feeds must be liquids, which can be coated on the rotating drum from a trough, or applied to its surface by rollers or by splashing. Product is scraped off by a blade. For the usual case of indoor installation, vapors are exhausted by a fan through a hood and ducts.

Temperature and exposure time (normally not more than 20 s) are regulated independently. Film thickness and solids concentration can also be controlled. Important applications are foods and products that are sticky when dry, are dusty when produced in other ways, or for some other reason cannot be handled by other dryers. One of the oldest dryer types, used on many different materials, it has been built in various configurations and feed systems. But practical temperatures are relatively low, and the massive, rigid design is essential to keep proper alignments and reduce the cost of maintenance. These factors have limited its uses to some extent.

3.3.4 Paddle dryers—high speed

The high-speed paddle dryer, shown in Fig. 3.2, is best suited to drying that is dominated by heat transfer rather than diffusion. It functions as either an indirect or a direct dryer, mostly as a combination of both. The majority of applications use some heated air, usually counterflow to the solids.

The jacketed shell can be zoned, but may not be heated when large airflows are used. The unheated agitator paddles have a tip speed that can be varied, but is typically 1000 to 2000 ft/min (5 to 10 m/s). They throw a thin layer of particles in a spiral around the inner surface of the shell, giving a high heat transfer rate. The action of the paddles breaks up many materials, which improves drying but adds fines to the product. When a product collector is needed, it is often installed

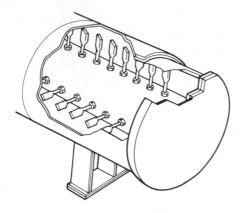

Figure 3.2 High-speed paddle dryer. *(Courtesy of Bepex Corp.)*

above the unit and the recovered powder is dropped back in. Some materials can be agglomerated while drying, usually requiring a spray of water and low-speed agitation.

A range of feed consistencies can be handled, including many wet and sticky pastes. A difficult feed may require backmixing—sometimes accomplished internally, other times needing zoned temperature control. A high heat transfer rate makes the unit suitable for wet feeds (even some slurries), which need a flow of air to carry off the vapor. Exposure times typically range from a few seconds to less than 20 min. For low final moistures, a secondary dryer with long exposure time (suited to diffusion drying) may have to be added.

3.3.5 Paddle dryers—low speed

The low-speed paddle dryer is intended for materials needing long exposure times (30 to 90 min). Thus it can function as a secondary dryer, often to a high-speed paddle dryer. T-bar-shaped, unheated agitators are mounted on a large-diameter shaft. These rotate slowly with the horizontal bar close to the jacketed shell, which may be either cylindrical or U-shaped. Solids are tumbled gently, avoiding severe attrition, but the heat transfer rate is low.

The main application is drying dominated by diffusion rather than heat transfer, removing firmly trapped moisture, usually at a low to moderate temperature. For some products just enough heat is added to the jacket to prevent heat loss. Suited to either continuous or batch mode, the units can be designed for atmospheric, pressure, or vacuum operation.

3.3.6 Rotary dryers

Of the various indirect rotary dryers, the most common is the steam tube rotary, named for its heating method; other means are used less often. Although the cost is relatively high, the different designs have important niches—in agriculture, in handling fine materials with little dust loss, and for operating as a closed system or under vacuum. The cylindrical shell and one to four concentric longitudinal rows of tubes rotate as a unit. Finned tubes are seldom used because most feeds tend to bridge over the fins.

Generally, a steam tube rotary uses no air or just a sweep, seldom more. Air and solids are in a kind of mixed flow, but the low airflow makes possible a modified plug flow. Materials can often be handled that require both heat transfer and diffusion dominated drying.

Conventional feeders can be used (screw conveyors reduce air in-leakage), but feed must be reasonably free-flowing to avoid fouling the tubes. Lifting flights are seldom used; instead the tubes cut through the material to provide some cascading, mixing, and breaking of lumps. The product is moved toward the discharge by the inclined rotation of the shell and by the head of material.

Sometimes external, hinged knockers are required. They fall against the shell as it rotates, and solids stuck inside are jarred loose. The discharge chutes also act as weirs to control the depth of solids loading in the shell. The loading, flight design, slope, and rotation speed of the shell (1 to 10 r/min) affect exposure time, which can be from 20 min to over 2 h. Loading of the shell, at 15 to 20 percent, is sometimes more critical than the exposure time.

Two variations on the steam tube rotary have rotating tube bundles and cylinders that are fixed, which reduces the problem of sealing against leaks. One operates with steam in the tubes and has a domed top for vapor removal. The other runs steam in the cylinder and feed in large-diameter tubes.

Another, very different type of indirect rotary is enclosed in a housing. With this design the outside of the rotating cylinder can be heated by combustion gases, occasionally by other means. Heat transfer is primarily by radiation from shell to solids. A tighter seal can be made with this rotary than with others. It can also handle very fine powders, produce special atmospheres, or operate at higher temperatures than is possible in the steam tube rotary.

3.3.7 Screw conveyor dryers

An outgrowth of the screw conveyor, this dryer has helical, hollow screws and a jacketed trough, both of which can be heated. It pushes the material gently with little intermixing, close to ideal plug flow. Heat transfer rates are low for plug-flow operation, but it is essential for applications that need the same exposure time for each particle.

Designs are with either single or multiple screws. In the latter case they may be separate for a more plug-oriented flow or intermeshed for better mixing. Lifter bars can be used for added agitation. Internal backmixing design can be added, but this offsets the plug flow and uniform heating. Exposure time and heat transfer rate are fixed to the speed of rotation.

4

Direct Dryers

4.1 Introduction

Direct dryers all have one common feature: air supplies the heat and carries off the evaporated moisture. But they differ greatly in how the solids move. The particles are at rest on tray dryers and they ride on conveyor dryers. They are stirred and tumbled in rotaries. They are thrown up like boiling liquids in fluid beds, get rapid transit in flash dryers, and are atomized into a mist in spray dryers. The action of the air varies from wafting over a static bed to making violent contact with every particle. Although system components are, in general, quite similar, the drying vessels are very different.

Drying vessels. Chambers and other housings for the major direct dryers are designed to suit the patterns of air and solids flow and to meet structural requirements. Operating pressures are usually just below 1 atmosphere. Thus except for rotary dryers, the chamber construction can be of sheet metal. This benefits cost and permits a more troublefree thermal expansion. In general, stainless steel provides corrosion resistance for parts that contact either the product or moist air.

Capacities. The most common direct dryers can be built in a range of capacities up to many tons of product per hour. Product rate is the ultimate concern to dryer users. But the heat load and its influence on dryer size and operating cost are more important to efficient performance and to profits. The heat load for direct dryers depends mainly on the evaporation rate, not the solids rate, except for very low moisture feeds.

Feeding. Ideally, feeds are prepared at the minimum moisture content that the dryer and feeder can accept. Feeders are available that can handle a range of concentrations from liquids to nearly dry solids. For wet solids feeders similar to those used by bed-type indirect dryers are used—screw conveyor, vibrating tray, rotary air lock, and others. But sticky or very wet solids have to be pretreated. Conveyor dryers, on the other hand, need more specialized types to spread the feed uniformly across their wide belts. Liquid feeds are sprayed by nozzles. Some spray dryers use nozzles, others use spinning discs.

4.2   Design Considerations

To achieve maximum production from the least investment, a dryer and its component auxiliaries must have the smallest volume. Thus the optimum design is the lowest volumetric airflow at the highest velocity (minimum exposure time) that meets the product requirements. The airflow, in turn, is determined by the required evaporation rate and the dryer inlet and outlet air temperatures, which govern how much moisture the air can accept. Except for low-moisture feeds, the amount of solids has little influence on a dryer's capacity. For a given dryer and airflow, evaporation rate and air temperature difference are the major influences on the production rate.

Air temperatures and flow of air and solids.   The best drying efficiency is at the highest air temperature at the dryer inlet and the lowest air temperature (or highest air moisture) at the outlet. The maximum at the inlet is limited by the strength-temperature properties of the metals. Another limit is set by the heat sensitivity of the solids and how long they are exposed to heat. At the low-temperature end the minimum is set by the product's moisture specification or by the possibility of moisture condensing at cool places in the ducts or equipment.

Cocurrent flow allows high inlet air temperatures, even for heat-sensitive materials, up to about 750°F (400°C) for some organic pigments. Because of the evaporative cooling, the air temperature can be much higher than the upper limit of the solids. Thus many heat-sensitive materials are dried at a lower cost by direct dryers than by indirect dryers, and at a much lower cost than by vacuum operation. But if a high-temperature operation is to be free of fires or scorching, product buildup must be avoided. A low-oxygen self-inertizing system (see Fig. 2.4) helps to avoid these dangers.

Other modes of flow are less common. Countercurrent flow exposes the outgoing dry product to the hot inlet air. Lower product moistures are possible, but drying action at the feed inlet is slower. The least common mode is mixed flow, used in some spray dryers, in which either the air or solids start in one direction and end in the other (see

Fig. 4.2*c* and *e*). The backmixed fluid-bed design is also a form of mixed flow of air and solids.

Air in-leakage. Because most direct dryers operate at a slight negative pressure, any leaks are inward. This minimizes the escape of powder. Old dryers often leak badly, but even new units cannot be made fully airtight. Leaks at the feeder (often the most serious) can be choked off by rotary air locks or screw feeders. Some systems, particularly those with high pressure drops, install two blowers in a push-pull arrangement. This results in a zero pressure point, which, when possible, is set where the equipment must be open at least some of the time. At such a null point, with a minimum of in-leakage, feed can be introduced, samples taken, or certain maintenance performed while the system is running.

Exposure time of solids in air. In the various direct dryers the air may pass over the solids or through them, or it may support or convey them. The solids may be in the air less than 1 s or over 4 h. This is variously referred to as exposure, residence, or dwell time. When the solid's exposure is hard to measure or calculate, the time used is that of the air instead of the solids. Another factor, sometimes overlooked, is the time the solids spend in the ducts and product collector. In flash and spray drying systems this may be a large percentage of their exposure to the heat.

In all but the static bed types most of the evaporation ends in a fraction of a second. The rest of the drying takes several seconds or minutes or longer, depending on the material and type of dryer. When the particles are not of uniform size—and this is the usual case—the smaller ones dry faster and are carried off more readily in the airstream. This may be either troublesome or beneficial. Fines ejected in the airstream have to be recaptured and mixed with the product, or returned to some point in the overall process, or disposed of. Fines are often overdried, and removal of fines alters the particle size distribution, which may or may not be desired.

Product properties. The properties of greatest concern for most products are moisture content, particle size, and bulk density. Depending on the material, other properties, such as color, taste, flavor, shape, and flow characteristics, may be even more critical to the final user.

But only moisture content is a major influence on other properties and conditions. There are several reasons for it.

1. Its evaporation keeps products from overheating.

2. It often binds particles into agglomerates. When the moisture content is too high, it can fuse them into a mass or block.
3. It is sometimes an influence on bulk density, ability to flow, degree of hydration, and color.
4. Always cheaper than the solid, it may be important from a marketing standpoint.
5. It is the final measure of the end of drying.

The air's direction of flow compared to the solids' also affects the properties, moisture content in particular. In cocurrent flow, if the heat is not dissipated evenly, the moisture will not protect all the solids from overheating. The relatively cool, moist air at the discharge is gentle to heat-sensitive products. In the opposite mode, countercurrent flow, the air enters at its highest temperature where the product is discharging. It more readily drives out the last traces of moisture, but at the risk of overheating. Its higher efficiency for low-moisture diffusion dominated drying may reduce the required exposure time, and thus vessel size. The reverse applies for high-evaporation heat transfer dominated drying.

Particle size is characterized as average size and as size distribution. Specially designed graph papers are used that render the data for most dried products as straight lines, making relationships clearer. The particle size of the product is generally the same as in the feed. But some solids are reduced or agglomerated either by abrasion or by the rigors of the drying process.

Particles formed by evaporation from droplets—in spray dryers most often, but not exclusively—are usually hollow spheres. With cocurrent flow at high inlet temperatures the effect is greatest because drying is fastest. Some materials form a skin that slows further drying, but causes the particles to balloon. This increases particle size and decreases bulk density. Some liquids evaporate so fast that droplets cannot form. The result is a very low density product with many tangled filaments. These hollow-sphere and filament effects can sometimes be modified by changing the drying temperatures or the feed solids concentration, or both. Or these effects may be altered by special additives in the feed, if permitted.

There are still other influences on product properties, especially in spray drying. Atomizer wheel designs and speeds affect the bulk density of particles and their size and size distribution. Bulk density is higher when the particle size distribution is nonuniform, because small particles fit into the voids. Thus the bulk density increases when particles break up. Agglomerates may break up if they have formed because binders that were present in the feed vaporized in the heat. Some particles, such as fragile crystals, shatter when they are agitated or are air-conveyed, as through ducts or cyclones.

Product properties are also affected by exposure time, system components, the nature of the original feed, and the temperature and humidity of the air, especially at the outlet. Of these only the air conditions can be controlled during operation. The others are fixed by the system design or by the nature of the feed.

Control. In most installations the dryer's two air temperatures are measured and used to control feed and fuel (or other heat sources). Generally the inlet air temperature is held constant by regulating the fuel rate, and the outlet air temperature is held constant by regulating the feed rate. Another method is for the feed rate to control the fuel rate. In this case the outlet air temperature is fixed and the inlet air temperature fluctuates.

Controlling the drying conditions aims to achieve the following:

1. Assure the right product moisture and other properties
2. Avoid condensation in the cooler parts of the dryer
3. Retain the same product moisture if operating conditions are changed

Air temperatures determined in the testing stage are set to obtain a specified product moisture. If the exposure is long enough, the moisture in the particles and that in the air reach an equilibrium. The time required is governed by the product's affinity for moisture—a function of bound moisture and particle size. Chemically bound, hydrated moisture can be removed by raising products to specified temperatures, which can be found in chemical handbooks.

The proof that air temperatures are set properly is gauged by measuring the product moisture (and other relevant properties). Moisture in the outlet air is also important, but hard to determine, and it is only needed if operating changes alter the product. Several measures are used to gauge the outlet air condition—wet-bulb temperature, dew point, percent relative humidity, and adiabatic saturation ratio (which is outlet air moisture over saturated air moisture at the same enthalpy). The first two can be measured. The latter two are calculated or read from a psychrometric chart, using the dry-bulb temperature and either the wet bulb or the dew point. There is no general agreement on which of these measures is best, indicating, perhaps, that none suits all dryers and all purposes. (See Sec. 5.2.8.)

For some products the final moisture is critical and must be controlled accurately. One method uses a continuous moisture analyzer to sense it and hold it to a fixed point by using it to regulate the outlet air temperature. This method is used, for example, in the spray drying of certain clays for tile and dinnerware. By holding the outlet temperature constant in this way, it is possible to maintain the consistently high moisture content needed to bind together the fine clay particles into agglomerates for pressing.

Dryers cannot usually control particle size, bulk density, and other properties, although spray dryers can to some extent. But equipment can be added to tailor particles to meet specifications. Grinding and screening equipment, for example, are in common use to reduce size or classify, or both. Also, when denser or less dusty products are needed, equipment can be added that compacts, crushes, and screens out the desired size, and recycles oversize and undersize. Agglomerating is another technique that reduces dustiness, and it also makes products that dissolve quickly.

Many drying systems do not run at full capacity at all times. Their ability to operate at reduced capacity is governed by the turndown ratio of their various elements. The most critical are the maximum turndown of fired heaters and the minimum air velocity needed to keep airborne solids from settling out. When the feed rate is reduced, the air velocity cannot be reduced to the same degree, if at all. Instead, for proper control the inlet air temperature must be reduced, especially in the case of flash dryers. If a system has a fired heater, the burner must be able to turn down enough to produce the lower temperature. If it has the usual control of feed rate by the outlet air temperature, the inlet air temperature has to be overridden manually when the capacity drops.

4.3 Specific Direct Dryer Designs

This section describes the functions, features, and operating methods of direct dryer types widely used in the process industries. But there are many other types, as well as variations of those covered here. The diversity of designs is great. A text of this kind can only comment on those considered to be the most prominent. More detailed descriptions on these and other direct dryers can be found in the references, in various other texts, and in manufacturers' literature.

4.3.1 Conveyor dryers

Also called band or apron dryer, this unit is well suited to dry materials of mostly large sizes, but when designed properly, it can also handle particulates. For these it uses a continuous metal screen or belt on which the materials ride. Its enclosure is a modular, insulated housing, rectangular in all views. For particulates the belt is usually perforated, so that heated air can pass through the solids, thus the name through-circulation dryer for this type. Some fine powders need a nonperforated belt, and the air impinges on, or flows across the material. After the solids drop off, the belt may be cleaned by brushing, washing, heating, or some combination of these.

Feeds most often are granular. Heavy pastes can be extruded into noodle-shaped pieces, about ¼ in (6 mm), but they must be dry enough to avoid reagglomerating into a mass. Feed has to cover perforated belts evenly, including at the edges; otherwise air bypasses and drying is nonuniform.

Although free-flowing materials can be fed with adjustable belt-width hoppers, most feeds require cross-conveyor or oscillating feeders. The material depth on the belt is typically 4 in (10 cm). Some slurries can be flowed onto solid belts, predried with impingement airflow, then broken up and dried to their final moisture content on a through-circulation type.

Air provides the main heating effect, but some heat enters the solids from the belt. Heaters and blowers inside the dryer heat the air and force it up through part of the bed, then down through the next section. Periodic reversal of airflow assures more uniform drying.

Recycling and reheating a portion of the air saves energy and gives good control of the air moisture. This helps to overcome shrinkage and case-hardening problems. Areas can be zoned to give more specific temperature regulation—a common practice is to cool after drying. Varying the air velocity into several zones is another control technique. Depending on the air velocity and the material, the air may pick up fines that foul the equipment. This can be reduced somewhat by using only downward airflow, but sometimes screens are needed to filter out fines.

Occasionally two or more conveyor dryers are staged in series to provide (1) delumping between stages, (2) different depths of solids, or (3) a better distribution of exposure time between heat transfer and diffusion drying modes, or between sticky and granular states. Another design has multiple conveyors in the same enclosure. This extends the exposure time when other conditions can be held constant.

There is no lag in the control of operating conditions, which are more easily measured compared with other dryers. This includes the moisture content of solids, wet- and dry-bulb temperatures, and the effects of shrinkage. The exposure time is typically from a few minutes to 4 h, but it can be much longer.

Treatment of the solids is very gentle, even in staged and multiple units, if solids are just dropped to each lower belt. Because solids are in strict plug flow and static, except between stages, the units are best suited to diffusion dominated drying.

4.3.2 Flash dryers

Although their drying vessels are simple vertical ducts, flash drying systems are often complex (see Fig. 2.3). Solids loading is limited by the need to convey wet feed, and air velocity must be kept above a minimum. This limits the exposure time to a few seconds, although the time for the heaviest particles may be extended. In general only unbound moisture can be removed. To adapt flash dryers to more uses, many diverse systems have been developed, including intricate feed arrangements, product recycle, internal and external milling, and combinations such as multistaging.

In ducts of practical lengths the air's residence time is less than 3 s. The exposure is so short because the minimum air velocity for conveying is typically 3500 to 5000 ft/min (18 to 25 m/s), depending on the size, shape, and density of the particles. Even higher velocities are sometimes needed to convey the wet feed vertically. In horizontal ducts product tends to settle out. Because of their air-conveying ability, flash dryers are also called pneumatic conveying dryers.

The interaction of particles and air flying together is complex. The heavier, slower particles lag behind. Many are randomly struck and broken up by fast-moving lighter ones. The collisions and the high duct velocities cause attrition of friable materials. These and any other fines are readily removed in the airstream. But wetter, heavier particles stay in longer, and because of the weight difference, they can be separated by centrifugal devices inside or outside the dryer and given further drying.

In most cases feed must be granular, and particles should not be so heavy that they drop out. Backmixing some recycled dry product into the fresh feed is a common technique. Consistency is needed to allow uniform dispersion into the dryer without sticking. Special rotary air-lock feeders or other dispersers reduce air in-leakage and feed contact with the walls. Feeds are occasionally sent into the dryer in an air-stream using an aspirating venturi feeder.

To assure uniform drying, feed may be deagglomerated or milled before or after entering the drying duct. Alternatively, a kicker mill, or sling, may be used to break up lumps and give the particles a boost. When these do not make the drying sufficiently uniform, a classifier can be installed in the dryer to remove fines (which dry faster) and to recycle the heavier fraction.

Even with all these options, to get the product dry enough, a second-stage dryer may have to be added. This may be another flash dryer, but for diffusion dominated drying it is more likely to be a fluid-bed or other dryer that offers long exposure. For special applications some flash dryers are designed to give longer exposure with expanded sections of the duct or a spiral path for the solids. The high-speed paddle dryer, normally considered an indirect dryer, can be used on some materials as a long-exposure-time flash dryer.

In addition to minimum air velocity and maximum solids loading, the air humidity, as measured by the wet-bulb or dew-point temperature, must not exceed a limit that would allow condensation to occur. Changing the feed rate or solids concentration can affect these parameters.

In general, flash dryers use more energy than others, because the air temperature difference, which is the driving force, is limited. The inlet air temperature is kept relatively low by feed contact with metal at the small-diameter inlet. The outlet is kept high to offset the short residence time that inhibits drying. Nevertheless, some applications permit very high inlet air temperatures, occasionally above 1200°F (649°C). The benefit of higher temperatures at the inlet, however, is partially offset by the need for somewhat higher temperatures at the outlet as well to avoid excessive humidity.

High inlet and low outlet temperatures provide optimum productivity and energy use. On the other hand, when a flash dryer is run at lower than design capacity, its airflow must still be at or near design velocity for conveying. To keep the proper outlet conditions, the inlet temperature must be lowered. This increases energy use per unit of evaporation or product.

4.3.3 Fluid-bed dryers

Fluid beds have a turbulent mixing action, much like a boiling liquid, which is ideal for drying and other processes. They have a wide range of possible designs, temperatures, and residence times. Of the principal direct dryers only the fluid beds use internal heaters and thus perform as indirect dryers. They can heat, dry, and cool in the same unit, and they can be used to coat or build up the size of particles.

In plan view fluid-bed chambers can be round, square, rectangular, or spiral. Their screens, grids, or dispersing plates are perforated to retain solids of many types and sizes and yet allow air (or some other gas) to pass through. It is important for many applications that the chamber be accessible for cleaning. Internal exchangers are particularly subject to fouling. Figure 4.1 shows two typical chamber configurations.

In operation air first flows into a plenum below the screen. Pressure

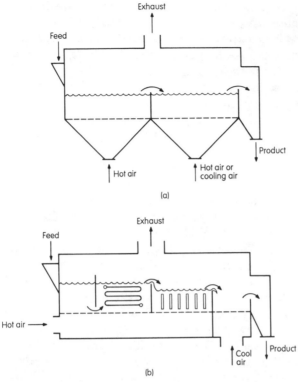

Figure 4.1 Fluid-bed drying chambers. (*a*) Two-zoned backmixed flow for optional heating, drying, or cooling. (*b*) Multizoned plug flow with internal tubular and plate heaters.

drop through the screen distributes the air evenly to the bed. The screen often has canopied perforations that direct the airflow and reduce the chance of solids falling through to the plenum. Air surrounds each particle, heats uniformly, makes other treatments possible, and gives some protection from attrition. But the violent action at long exposures breaks up very friable materials.

Airflow is by two blowers—forced draft at the inlet and induced draft at the outlet (or before the collectors), giving a push-pull effect, with zero pressure at or near the feed entry. Air velocity has to be above incipient fluidization. But it must not be so high as to elutriate too much product. Air velocities typically range between 100 and 400 ft/min (0.5 to 2.0 m/s). To reduce the air velocity over the fluidized bed, the upper portion of some chambers is enlarged. With an adequate freeboard disengaging section above the bed only fines are elutriated and the desirable fractions are retained.

Chambers are designed for either backmixed or plug flow. The bed area and depth give the necessary residence time as found by testing. Backmixed beds have round or square chambers. Because evaporation is rapid, these are best suited to heat transfer dominated drying, especially of surface moisture. Vigorous mixing and rapid evaporation make the average bed temperature essentially the same as that of the product. But the mixing action extends the residence time distribution between particles.

Plug flow can be simulated by the use of baffles. For the most diffusion-dominated, low-product-moisture jobs, a flow path with much greater length than width is needed. This can be a rectangular, zigzag, or serpentine design. Combination mixed and plug flow units can dry in both heat transfer and diffusion modes.

The air is heated by conventional means, external to the dryer, but it can be brought into the chamber through several zones, designed for different temperatures or velocities, as required by the condition of the solids. This gives better control and allows separate heating, drying, and cooling.

The indirect internal heaters used in both backmixed and plug flow fluid beds can be tubular or plate type. They add heat to the solids (usually by steam) and can supply over two-thirds of the heat load. But exhaust humidity is increased, and this limits applications. Maximum exchanger surface minimizes the required airflow, energy use, and equipment required for heating air and separating fines.

Feeds are most often wet solids, but spraying of liquid feeds into a bed of dry material is not uncommon. Feeding is usually into the top of the bed toward the center. Some sticky feeds can be handled because the active bed is a true backmixer. Internal heaters are baffled off or located out of the first section of zoned chambers to avoid fouling their surfaces when handling sticky or otherwise difficult feeds. But the internal heating effect can be extended back to the first zone by flow under the baffle.

In backmixed dryers the usual control method is to regulate the inlet air temperature by the bed temperature, which is essentially the same as the product temperature. With multiple zones, the weir height in each governs the residence time, which is often 10 to 25 min, but may range from 1 to 60 min. The wet-bulb or dew-point temperature must be held below a value determined in the testing stage, known to be acceptable.

The pressure drop in a fluid bed is typically 50 to 100 percent higher than in other direct dryers. This takes more power, but is still a minor part of the total energy cost. Deeper beds, thus higher pressure drops, are needed for longer residence times and for sticky feeds. If the particle size distribution is broad, fines are swept out. In some applications this separation is desired, and in others the fines can be captured by internal or external collectors and returned to the bed.

In spite of the versatility, not all materials can be fluidized. Particles should be of uniform shape in the size range of 100 μm to 10 mm with narrow distribution. Particles must not be excessively sticky in the feed or product, and they must be resistant to either lumping or abrasion. In addition, the material must not be too dense or cause uneven air distribution, which would result in slugging or channeling.

It is often possible, however, to handle such materials in a vibrating fluid-bed dryer. These shallow bed units fluidize without air by oscillating to throw the material up off the bed. Thus airflow is needed only for heating. More difficult materials can be handled using less power with much less attrition and less elutriation of fines. A frequent use is secondary drying (such as after a spray dryer), in which hot air through the screen is reduced to the minimum for economically removing final traces of moisture.

Spouted bed dryers are like fluid beds, but their action is somewhat different. The feed is conveyed by the hot air into the center of the vertical cylindrical vessel at a velocity that throws the solids upward through the mass of material that is moving back down outside the spout. Main uses are for products such as grains and coal, with particle sizes too large to be fluidized at practical air velocities.

4.3.4 Rotary dryers—direct

Direct rotary dryers, like indirect steam tube rotaries, have a sloped, rotating cylinder. Often called the workhorses of the industry, they have the lowest total operating cost for many materials, notably those dried in bulk and at high temperatures. Wide particle size distributions can be handled if the fines dry quickly and are airborne out, while the heavier particles, moved mechanically, stay in longer. Under some circumstances they can simulate flash, spray, and other dryers. The mode of operation is mixed, rather than plug flow, and is best for applications dominated by heat transfer rather than diffusion.

All heat is supplied by hot air (sometimes flue gases can be used), with flow countercurrent or parallel to the solids. Most drying takes place as the solids cascade off lifting flights and the air passes through them. But solids are also heated partly by the shell. For products that must dry in a lower oxygen content, the drying gas is self-inertized by recycling part of the exhaust back through the fired heater.

Preferred feeders are the vibratory and mixer types and, when needed to help seal against leaks, the screw conveyor. With the right combination of features, even some liquids and sludges can be sprayed into a bed of dry material and dried successfully. But for proper action through the dryer, the solids must be granular. Wet or sticky feeds can be backmixed externally with dry product, or fed into a bed of dry material, dammed at the inlet. Devices that aid in handling wet feeds include internal hanging or scraping chains and external hinged knockers.

Solids are moved through the shell by several effects—airflow, head of material, flight design, and speed of rotation and slope of the shell (which slide and cascade the solids). Countercurrent airflow retards the solid's progress; cocurrent flow advances it, but may carry out a greater portion of fines.

Past experience with similar materials governs the spacing of lifting flights and their design (influenced by the product's angle of repose). Typical speeds of rotation range from 1 to 10 r/min, but may run as high as 30 r/min. The percent loading of the shell, also determined by experience, is 9 to 15 percent, occasionally higher. Both loading and exposure time, which ranges from a few minutes to as long as 4 h, are critical to desired product results and clean operation.

Some variations of the direct rotary are designed to hold the solids longer at the end of the drying cycle. This extends the exposure where it is most needed. Diffusion dominated drying is thus improved for the effective drying of materials with bound moisture.

Perceived disadvantages of the rotaries stem partly from their versatility and numerous applications. At low capacities and low temperatures they are not as cost-effective. Radiant heat losses are high when temperatures are high, especially in counterflow, because insulating the hot end could cause metal failure (as could feed cutoff). Insulating the discharge end, however, is needed for low-temperature jobs to avoid condensation. External knockers or internal chains are needed to unstick some materials from the shell, but the noise makes them unacceptable in some locations.

4.3.5 Spray dryers

Spray dryers are widely used on inorganic materials as well as organics, such as foods, their original application. The main reasons for their acceptance are that (1) high inlet temperatures are possible, (2) one or more processing steps are often eliminated, and (3) for many materials the required size, shape, density, or other property cannot be made any other way. Because spray drying is limited to liquid feed, the evaporation load is high. Thus energy and airflow requirements are also high. Nevertheless, spray drying is used on a wide range of materials, and many designs have evolved.

Atomization.   Feeds must be sprayed in a uniform pattern at a uniform rate into a uniform nonpulsating flow of heated air. Not all liquids can be spray-dried, notably those that are tacky when dried and stick to equipment surfaces. But some materials that go through a sticky phase during the airborne stage can only be spray-dried. Any liquids that can be pumped and sprayed into fine drops are worth testing as long as they are not too viscous—solutions, slurries, emulsions, and others, even many thixotropic pastes. The atomizing capability is determined largely by viscosity and surface tension. Properties may have to be modified by heating or by diluting, which is not a desirable option because the evaporation rate determines dryer size and operating cost.

Feeds are atomized using either nozzle or spinning discs. Both types have about the same ability to produce a specific droplet size and uniform size distribution. In typical chambers the average size may be as small as 10 μm or as large as 150 μm. Larger sizes can only be made in large-diameter or tall chambers. Small sizes cannot be made at high capacities, except at high atomizing power.

In general, nozzles are more economical, and they can spray in any direction. Except when high-pressure pumps are needed, maintenance is much simpler. But if feeds are not uniform, the nozzles tend to plug and cannot always operate uniformly if solids content or viscosity vary. In addition, abrasion enlarges the nozzle opening, changing the character of the spray. Multiple nozzles are needed for high production rates, and frequent cleaning and replacing are normal requirements. Thus their versatility and capacity are limited.

Single-fluid, or pressure, nozzles use the pressure of the liquid itself as the energy source to break the liquid into droplets, and uniformity is about equal to that for spinning discs. But the feed rate must be held constant to maintain product quality. For very viscous feeds high pressures are used.

Two-fluid nozzles, on the other hand, get their energy from compressed air, usually about 60 to 100 lb/in^2 (400 to 700 kPa). This is simple and convenient, but too expensive for high capacities. These nozzles accommodate feed-rate changes better. However, they give a broader particle size range, which sometimes is desired but often not. Designs that mix the air and feed internally are the most efficient. Those that mix externally resist wear, and thus are useful on ceramics and other abrasive feeds, which quickly erode other designs.

Spinning discs, also called atomizer wheels, can be used on virtually all applications—for abrasive feeds, for general use, and for special jobs such as producing uniform or fine droplets. They are the preferred choice for high capacities. Spray can only be horizontal, so they are not used in tower designs. Compared to nozzles, discs have higher capacities, plug less, and are affected less by abrasion and by changes in feed properties and feed rates.

Required atomizer speeds are typically 5000 to 20,000 r/min, depending on the droplet size needed and the disc diameter, which can be 4 to 12 in (10 to 30 cm). Drive mechanisms for spinning discs are far more complex and costly than nozzles (except for nozzle systems that require staged high-pressure pumps). These drives are either standard motors (with speed increasers or belt drives) or direct-drive "spray machines" built around high-speed motors operated by high-frequency current. They require accurate balancing and skilled periodic maintenance.

Such considerations are important, but the choice between the two types is often based on other factors. These include the required feed rate and particle size and the kind of drying chamber needed—or what the preferred supplier has to offer. The type of atomization has only a minor effect on the product, but it determines the chamber shape to some extent. Spinning discs can only be used in chambers large enough in diameter to accept their horizontal sprays. Nozzles can be used in any chamber, but performance is best in towers.

Chambers. There are three categories of chamber geometry—short cylinders with height-to-diameter ratios of 1.0 or less, tall cylinders with ratios greater than about 2.0, and a rectangular box design. Figure 4.2 shows the chamber geometries and the spray and airflows of seven common designs. Those with the widest range of applications are the conical and tower types with cocurrent air-spray patterns. Table 4.1 lists the chambers shown in Fig. 4.2 and gives their airflow modes, atomization types, and principal uses.

Figure 4.2*a* is the most versatile and commonly used design. Atomization can be by spinning disc or by one or more nozzles, for which the chamber might be somewhat taller. The product leaves at the bottom. Air is exhausted either at the bottom with the product or further up the cone, in which case it only carries out the fines.

Exposure time. The average residence time of the air ranges from 5 to 30 s in most spray dryers, but it may be as high as 60 s in tall towers. Certain products require special designs. Those used for drying coffee are similar to that of Fig. 4.2*d*, except that they have slow parallel nonturbulent airflow and a large-diameter product-air separating zone from the top of which the air is taken out.

To make large particles, a long exposure is needed. To make large particles at low capacities, a small-diameter relatively tall chamber is used. Exposure of the solids, the real measure of drying time, is lengthened in the mixed-flow design shown in Fig. 4.2*e*. Spraying up-

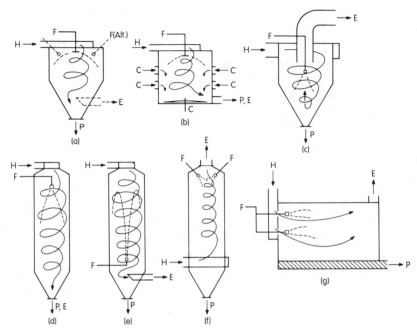

Figure 4.2 Spray-drying chambers. F—feed; H—hot air; C—cooling air; P—product; E—exhaust. (a) Conical. (b) Cylindrical. (c) Mixed flow. (d) Cocurrent flow tower. (e) Mixed-flow tower. (f) Counterflow tower. (g) Box.

TABLE 4.1   Spray Dryer Applications

Part of Fig. 4.2	Chamber type	Airflow mode	Atomization type	Use
a	Conical	Cocurrent	Disc or nozzle(s)	General; most widely used
b	Cylindrical	Cocurrent	Disc	Heat-sensitive products
c	Cyclone	Mixed	Nozzle	Limited; high-density products
d	Tower	Cocurrent	Nozzle(s)	General
e	Tower	Mixed	Nozzle	Low capacity; inorganics, ceramics
f	Tower	Countercurrent	Nozzle(s)	High density; detergents
g	Box	Cocurrent	Nozzle(s)	Dairy, food products

ward gives a maximum travel path, without a costly large-diameter chamber.

Combinations.   Spray-dried particles are small and usually of low bulk density. Some are dusty and some are slow to disperse in liquid. These disadvantages can to some extent be adjusted by varying the feed solids or air temperatures, but added equipment may be needed. One method is to add a fluid-bed secondary dryer to agglomerate the particles. Products treated in this way are said to be "instantized"—they disperse rapidly, if not instantly.

One integrated drying and agglomerating system sprays the feed into a chamber mounted directly over a mesh belt. Partially dry solids form a porous mat on the belt through which the drying air is filtered. The mat can be conveyed through other zones for additional processing. This can be added retention time, more drying, or cooling. The product, fast dissolving and relatively free of fines, is then classified to meet the needed size specification.

4.3.6   Tray dryers

Traditional tray dryers operate in a batch mode, although very large units can be semicontinuous. Depending on the size and specific design, they are referred to as tray, truck, or tunnel dryers. The equipment cost is the lowest of all the principal dryers, but with manual loading and unloading of the trays, the operating cost is the highest.

The continuous rotating tray, or plate, dryer largely overcomes the static tray dryer's most serious disadvantages—high labor and space needs, uneven heating, and long exposure time. The trays are circular, and the top tray receives the feed from a conventional feeder. After one revolution the material is wiped off and drops to the next lower tray. This process continues, and on each tray the material is leveled and raked by stationary arms, so that the discharged product is dried more uniformly.

Hot air from centrally mounted blowers heats both the top of the material and the trays, which in turn heat the bottom of the material. Air can be recycled and its flow rate varied, as can tray speed and material thickness. Recirculation of a portion of the exhaust improves energy efficiency, especially at low air moisture. The resulting dried products are not always particulates, but may be small lumps or cakes that have to be ground and screened.

Calculation Methods

Equations, charts, tables, and rules of thumb in some combination, tempered with experience, are needed to transform data and ideas to hardware. For designs to be successful they have to be based on logic, or else on empirical relations and procedures that are firmly grounded in experimental results or proven practices.

Much of the calculating needed for dryers centers around the heat needed for evaporation. Also considered are feed and product temperatures, leaks, collector efficiencies, heat losses, heat source, operating pressure (usually determined by elevation), and heat of crystallization, reaction, or other such effects. Calculations alone, however, can seldom complete a design. Experience and test trials play major roles in most aspects of drying practice.

The units in this chapter are in the U.S. customary system, except that data from the literature are generally left as given and converted as required. Conversions to SI (système international) and other systems are given in App. A.

5.1 Indirect Dryers

5.1.1 Calculation procedure

Heating, constant-rate drying, and falling-rate drying occur in most indirect drying operations; cooling is another possible step. The heating and cooling surfaces for each can be figured separately, but often one operating mode dominates to such an extent that its average heat transfer rate can be used as the design basis. Heat transfer rates are not computed as the sum of resistances as they are for heat exchangers. Instead they are found from test data—heat loads, temperatures, and the known heating surface of the test dryer. Sometimes the rate and the temperature difference are combined into a single term.

The calculation procedure is simpler if the heat rate is relatively constant throughout the length of the dryer. On the other hand, when the heat rate is high at the start of drying and low at the outlet, the requirements for both heating and agitating are very different at the two ends. A single dryer with several heating zones might be adequate, but often two dryers in series are better. The first dryer is calculated by the heat transfer rate, the second by exposure time, with data for both determined by testing.

The example given in this section shows how the heating surface can be calculated for an indirect dryer using heating, constant-rate drying, and falling-rate drying. Heat transfer dominates throughout, but in each zone the heat load, heat transfer rate, and temperatures are different, requiring separate calculations of the surface area. Each is adjusted for the loading. The heating medium is steam; only its saturated temperature is used for calculating the temperature difference. Superheat contains little useful heat and has poor heat transfer properties.

Nomenclature and units

A	Heat transfer surface area, ft^2
C_l, C_s	Specific heat of liquid, solid, Btu/(lb · °F)
L	Loading or coverage of heat transfer area, %/100
L_v	Latent heat of vaporization of liquid, Btu/lb
Q_l, Q_s, Q_e	Heat load of liquid, solid, evaporation, Btu/h
Q_t	Total heat load, Btu/h
W_l, W_s, W_e	Flow rate of liquid, solid, evaporation, lb/h
U	Heat transfer rate, Btu/(h · °F · ft^2)
T_i, T_o	Temperature of heating medium, inlet and outlet, °F
t_i, t_o	Temperature of liquid or solid, inlet and outlet, °F
T_m	Log mean temperature difference, °F

Equations.   The equations are identical for each zone. The log mean temperature difference, in °F, is

$$T_m = \frac{(T_i - t_o) - (T_o - t_i)}{\ln[(T_i - t_o)/(T_o - t_i)]} \tag{5.1}$$

The heat loads, in Btu/h, are for liquids,

$$Q_l = W_l C_l (t_o - t_i) \tag{5.2}$$

for solids,

$$Q_s = W_s C_s (t_o - t_i) \qquad (5.3)$$

for evaporation,

$$Q_e = W_e L_v \qquad (5.4)$$

The total heat load is

$$Q_t = Q_l + Q_s + Q_e \qquad (5.5)$$

The required surface area, in ft^2, is

$$A = \frac{Q_t}{T_m UL} \qquad (5.6)$$

5.1.2 Example calculations

Data for the example were chosen to show each operating zone that contributes to the total surface. The data, including test results, are given in Table 5.1. The calculations are shown in Table 5.2.

When backmixing is needed, the added feed volume mandates a like volume increase in the vessel. It may also require increased heating surface because the wet particles make less frequent contact with it.

TABLE 5.1   Data for Calculation Example of Indirect Dryer

Feed rate	1000 lb/h
Feed moisture	20%
Feed temperature	60°F
Product moisture	5%
Product temperature	260°F
Solids specific heat	0.4 Btu/(lb · °F)
Water specific heat	1.0 Btu/(lb · °F)
Heating medium	Steam
Temperature	338°F
Enthalpy	880.8 Btu/lb
Thermal data for moisture and solids	
Evaporation enthalpy	970.3 Btu/lb
Water enthalpy at 212°F	180.2 Btu/lb
Water vapor enthalpy, averaged 212–260°F (1150.5 + 1167.4)/2	1159.0 Btu/lb
Solids temperature in constant rate	212°F
Product moisture at start of falling rate	10%

	Zone		
	1	2	3
Heat transfer rate, Btu/(h · ft^2 · °F)	25	40	15
Surface loading, %	100	80	60

TABLE 5.2 Calculations

			Flow rates, lb/h
Dry solids	1000 (1 − 0.20)	=	800
Product	800/(1 − 0.05)	=	842.1
Moisture at change to falling rate	$\dfrac{800 \times 0.10}{1 - 0.1}$	=	88.9
Moisture in final product	842.1 × 0.05	=	42.1
Evaporation			
Total	1000 − 842.1	=	157.9
Falling-rate zone	88.9 − 42.1	=	46.8
Constant-rate zone	157.9 − 46.8	=	111.1

	Zone				
	1		2	3	
Operation	Heating		Constant-rate drying	Falling-rate drying	
Temperatures, °F					
Steam	338	338	338	338	338
Solids	60	212	212	212	260
	278	126	126	126	78
Log mean difference	192.1		126.0	100.1	
Heat loads, Btu/h					
Solids	800(212 − 60)0.4			800(260 − 212)0.4	
Liquid	200(212 − 60)1.0			42.1(260 − 212)1.0	
Evaporation			111.1 × 970.3	46.8(1159 − 180.2)	
Zone totals	79,040		107,800	63,189	
Surface area, ft² (adjusted for loading)	$\dfrac{79,040}{192.1 \times 25 \times 1.0}$ = 16.40		$\dfrac{107,800}{126 \times 40 \times 0.8}$ = 26.74	$\dfrac{63,189}{100.1 \times 15 \times 0.6}$ = 70.14	
Total surface required				113.3 ft²	
Total heat load				250,029 Btu/h	
Steam rate	$\dfrac{250,029}{880.8}$	=		283.9 lb/h	

But that effect is offset to some extent by the direct contact of warm, dry particles with the wet feed.

5.2 Direct Dryers

5.2.1 Introduction

Calculations for direct dryers are very different from those for indirect dryers because they involve the psychrometric relations of moist air. The volumetric airflow rate has to be computed because it is the major influence on the size and cost of the drying vessel and auxiliary equipment. Also, the airflow rate and heat load are the main influences on operating costs.

This procedure includes all the common heat and mass inputs and outputs, collector efficiencies, system pressure, and leaks. It shows the calculation method for the heat load and for the flow rates and properties of air at five dryer locations, as well as the saturation conditions at the dryer outlet. The necessary properties of air and water are calculated from equations, either directly or using data from App. B.

The vessel diameter is calculated after the air conditions have been computed. Depending on the dryer type, various other routines are needed, such as adjusting the airflow rate or velocity to meet specified conditions of air humidity, exposure time, and solids-to-air loading.

The procedure is arranged for manual calculations or for setting it into a computer program. Adapting it to the latter speeds multiple runs and helps avoid mistakes. Desired conditions, such as for field or pilot plant tests, can be found by trial-and-error reruns. Calculating manually, with or without psychrometric charts, on the other hand is error-prone and time-consuming. Charts, in particular, cannot include all the effects, although they give a good picture of the process and are practical for estimates or for a few simple calculations.

5.2.2 Definition of variables

The variables have been devised to keep the equations recognizable and to simplify converting them into computer programs. The subscripts of arrays signify the dryer stations, except for air enthalpies K, which have 21 elements to provide enough values for calculating from 0 to 2000°F.

Figure 5.1 shows the five station numbers for a spray dryer, but they apply to most dryers using air (or another gas) as the heating medium. Station 1 is the entry point of air—usually ambient—into the heater. After heating, the air is at station 2 conditions. At station 3 drying is completed. Station 4 allows for entry of a leak or auxiliary air at some point after station 3. The station 4 air may be ambient or conditioned air, such as introduced into some dryers to cool sticky products before discharge, or it may be added to the airstream to protect bag filters or other equipment. Station 3 and 4 airflows combine to form the air at station 5. Thus if there is no auxiliary air or leak, stations 3 and 5 are the same.

For convenience in calculating, much of the procedure is on the basis of a pound of dry air (DA), as in Btu/lb DA.

Nomenclature and units. Arrays are shown with subscript i, which designates one of the five dryer stations, or with subscript j, which is used

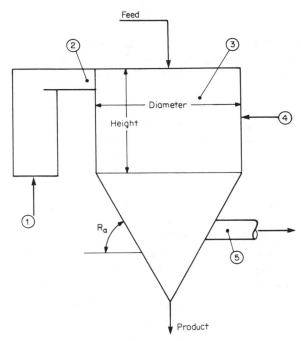

Feed

Diameter

Height

R_a

Product

Figure 5.1   Dryer stations.

for air enthalpies. With few exceptions, units are in the U.S. customary system—ft, lb, Btu, °F. Exposure times are in seconds. Rates are per hour, except for air rates, which are in lb/min and ft³/min.

A_i	Airflow rate, lb/min
A_r	Surface area of drying vessel, ft²
A_s	Adiabatic saturation ratio (ASR), %
C_i	Specific heat of dry air, Btu/(lb · °F)
C_p	Specific heat of dry air, fixed value, Btu/(lb · °F)
C_s	Specific heat of solid, Btu/(lb · °F)
C_a	Coefficient for Eq. (5.75), = 3.2438
C_b	Coefficient for Eq. (5.75), = 5.8683 × 10⁻³
C_c	Coefficient for Eq. (5.75), = 1.1702 × 10⁻⁸
C_d	Coefficient for Eq. (5.75), = 2.1878 × 10⁻³
C_n	Coefficient for Eq. (5.75), = 2.3026 (= ln 10)
C_g	Conversion factor from lb/h to gr/ft³ (grains; units are inconsistent)
D_s	Dry solids rate, lb/h
D_m	Vessel diameter, ft
D_e	Vessel diameter, estimated, ft

D_y	Air-vapor density at station 5, lb/ft^3
E_a, E_b, E_c	Collector efficiencies, % (drying vessel may be first collector)
E_t	Net collector efficiency, %
E_l	Plant elevation above sea level, ft
E_v	Evaporation rate in dryer, lb/h
E_s	Evaporation rate in scrubber, lb/h
E_x	Exposure time of air in drying vessel, s
F_i	Volumetric airflow rate, actual (ACFM), ft^3/min
F_d	Feed rate, lb/h
F_s	Feed solids content, wet basis, %
G_a, G_b, G_c	Dust loadings into collectors, gr/ft^3
G_d	Dust loading out of last collector, gr/ft^3
H_i	Air-vapor enthalpy, Btu/lb DA
H_x	Air-vapor enthalpy for T_w calculation, Btu/lb DA
H_c	Heat of crystallization, Btu/lb product
H_d	Height-to-diameter ratio of cylindrical vessel
H_l	Heat loss, % of station 2 enthalpy H_2
H_t	Heat loss heat transfer rate, Btu/(h · °F · ft^2)
H_v	Gross heating value of fuel, Btu/lb fuel
K_j	Dry-air enthalpy, Btu/lb DA (see App. B)
L_k	Leak into dryer, % of A_2
L_p	Auxiliary airflow into dryer, % of A_2
L_t	Total of leak plus auxiliary airflow, % of A_2
M_i	Air moisture, absolute humidity, lb/lb DA
M_s	Air moisture at saturation, lb/lb DA
M_f	Combustion moisture, lb/lb total DA
M_x	Combustion moisture, lb/lb stoichiometric air
M_h	Molecular weight of fuel
M_a	Molecular weight of air, = 28.97
M_v	Molecular weight of water, = 18.02
N_c	Number of carbon atoms in chemical formula of fuel
N_h	Number of hydrogen atoms in chemical formula of fuel
P_a, P_b, P_c	Product rate from each collector, lb/h
P_n	Net product rate from collectors (= P_r), lb/h
P_r	Specified product rate, lb/h
P_g	Gross product rate including loss, lb/h
P_l	Product loss from collectors, lb/h
P_m	Product moisture content, wet basis, %

P_t	Pressure, lb/in^2
P_x	Pressure correction for elevation
P_e	Exponent for e in Eq. (5.76)
P_p	Partial pressure of water, lb/in^2
Q_l	Heat loss, Btu/h
Q_t	Heat load, Btu/h
R_c	Weight ratio of combustion water vapor to fuel burned
RH	Relative humidity, %
R_a	External angle of vessel cone, degrees (must be in radians for some computer languages)
S_p	Flow rate of solids in product, lb/h
S_l	Slope of combustion moisture, lb moisture/(°F · lb DA) (see App. B for values for various fuels)
T_i	Air temperature, °F
T_d	Air temperature, dew point, °F
T_s	Air temperature, saturation, °F
T_w	Air temperature, wet bulb, °F
T_f	Feed temperature, °F
T_p	Product temperature, °F
T_q	Product temperature, intermediate value, °F
T_k	Air temperature, K
T_x	Temperature difference, $T_{water,\ critical} - T_{air}$, K
V_i	Humid volume, ft^3 dry gas and its vapor/lb DA
V_f	Volume factor for vessel with one conical end or both ends flat, $= V_u/D_m^3$
V_u	Vessel volume, ft^3
W_i	Vapor enthalpy for water, Btu/lb
W_c	Critical pressure of water, $= 218.17$ atm
W_f	Flow rate of water in feed, lb/h
W_p	Flow rate of water in product, lb/h
Z_b	Lumped heat term for solids and liquid
Z_h	Lumped heat term for air
Z_m	Lumped heat term for moisture
$int(n)$	integer portion of argument n

5.2.3 Input values

Input data for a calculation come from several sources—test results, conditions at the plant location, flow rates, temperatures, and other specifications for the application. Wrong input values are the main cause of bad results, so input data should be checked carefully. This is especially true for computer calculations, for which there is little opportunity to see intermediate values and how they were obtained.

Another help in curbing computer errors is to keep the input process simple. One technique is to use zero entries as flags. For any specific item zero signals one of these actions: (1) the value will remain zero, (2) a value will be calculated, or (3) a default will be assigned. Examples of defaults are $E_l = 1000$, $T_1 = 60$, $M_1 = 0.01$, $C_s = 0.4$, and $T_f = 60$. When collector efficiencies E_a, E_b, and E_c default to 100, 0, and 0, all product is assumed to be collected and none lost. C_p, H_l, T_p, and P_t can default to calculated values when zeros are entered.

If all the defaults listed are triggered, the only mandatory entries are T_2, T_5, P_r, P_m, F_s, and E_x. The other eight can remain zero, although for most jobs some have to be specified. Exceptions are M_h and H_v, which, if not needed, should be set to 1.0 to avoid division by zero in Eqs. (5.31) and (5.32).

Details needed for some input items are described in the following subsections. Leaks, product–air humidity relations, and calculations for other parts of the system are given in Sec. 5.2.8. Combustion data for commonly used fuels are listed in App. B.

Operating pressure. Most direct dryers operate slightly below atmospheric pressure to avoid powder leaks. In the procedure, the system pressure could be entered directly when it is other than atmospheric, but normally it is calculated from the elevation, which varies widely.

Many industries are in coastal areas below 400-ft elevation, but inland areas are often 10 to 15 times as high. Much of the United States is at about 1000 ft (Jorgensen, 1983). Comparing 5000-ft elevation against sea level for the same duty, the required volumetric airflow is 20.2 percent greater and the dew point is about 6°F less.

Feed solids or moisture content. The feed solids content can fluctuate daily, or even hourly, and thus have a major effect on both airflow and heat load. In some plants it varies more than 15 percent from one day to the next. Certain large installations would benefit by having a full-time technician monitor it.

Supply air moisture and temperature. The air supplied to the heater is normally ambient. In very humid areas its moisture content can be greater than 0.02 lb/lb DA (as for 90°F dry bulb, 80°F wet bulb). High ambient moisture raises the outlet air moisture accordingly, and for some products this requires raising the outlet air temperature. Higher ambient moisture also influences both fuel consumption and moisture formed by fired heaters. To illustrate its effect on the outlet humidity at moderate drying temperatures, raising the inlet air moisture from

0.01 to 0.02 is shown to raise the dew point at the dryer outlet by about 8°F.

Wide swings in the supply air temperature also affect results, particularly heat and power loads. Average ambient conditions can be used to compute yield and requirements for heat and power needs, but weather extremes should be used for sizing the equipment. For example, the sizing of heaters should be based on the lowest expected temperatures. The same is true for blower motors, which may overload during cold weather start-ups unless the airflow can be reliably dampered back. When airflow rates are critical, they should be based on design standards for summer air conditions (Jorgensen, 1983).

Heat source. A direct-fired heater adds water vapor to the air by combustion of the fuel. (Calculation is detailed in Sec. 5.2.8.) The moisture added depends on the fuel's molecular weight, the number of hydrogen atoms in its chemical formula, and its gross heating value. The data needed for calculating the added moisture are given in App. B for liquid and gaseous fuels commonly used for drying. The added moisture can also be computed using a shortcut method for which data are given in App. B as well. In either case, a correction may be needed for moisture in the supply air.

Waste gases. Flue gases or other hot waste gases may offer economical heat sources, but their moisture content is often too high to allow the desired product moisture to be reached. Another problem is that condensation in the system may frustrate continuous operation. For a calculation using waste gas, use its temperature as both T_1 and T_2 and its moisture as M_1 and M_2.

Heat loss and safety factors. Radiant and convective heat losses from a dryer are functions of the vessel's outer surface area and the temperature difference and heat transfer rate across it. For well-insulated cocurrent dryers with large surface areas the heat loss is only 0.2 to 1.0 percent of the heat load because most of the inner surface is close to the air outlet temperature. But it may exceed 1.0 percent at higher surface temperatures, from vessels less than 3 ft in diameter, and from those having a high ratio of height to diameter.

Heat loss from the surface of a drying vessel dissipates some of the energy intended for drying. The calculation procedure accounts for this by reducing the air enthalpy at station 2. This is a convenience in deriving and using the equations. It would be more correct to reduce the difference in enthalpies between stations 1 and 2 ($H_2 - H_1$), but the error is small and on the safe side.

The total heat loss through the dryer thickness, including insulation, is greater at higher surface areas and temperature differences.

But as a percent of the total heat load, the loss is lower at higher surface areas because the ratio of surface area to flow rate is lower. The total heat required, however, must include losses from the heater shell and inlet ducts.

The units for heat transfer rate are Btu/(h · °F · ft^2) and the rate is the reciprocal of the sum of resistances through the insulation, metal, and air films inside and outside the dryer. The insulation resistance is about 95 percent of the total. For a dryer with 2.0 in of mineral fiber blanket insulation and inside and outside surface temperatures of 200°F and 80°F, respectively, the heat loss is about 13 Btu/(h · ft^2), which is a rate of 0.11 Btu/(h · °F · ft^2). On the other hand, if that same dryer were uninsulated, the heat transfer rate would be about 2.3, so the loss would be 20 times greater.

The effect of heat loss varies too much for it to be useful as a safety factor. To illustrate, a 5 percent heat loss increases the airflow and heat load typically by 6 to 20 percent, even more in some cases. The effect is best seen by plotting the data on a psychrometric chart. It is more practical, after the airflow has been calculated, to add proper safety factors to the heater, blower, and other components that could become bottlenecks.

Product temperature. Although preferable to making an estimate, measuring the product temperature is difficult. Air in the sample affects readings, as does any remaining unbound moisture. With cocurrent airflow the temperature of the product can be higher than that of the air at the outlet. This is more likely to occur when the exposure time is short, moisture is only at the surface, and the inlet air temperature is high.

In the calculation procedure the product temperature can be entered if known. Otherwise empirical equations compute it as a function of the dryer outlet wet- and dry-bulb temperatures. That value is then adjusted by the final product moisture. The result is an approximation, but its effect is significant only when the evaporation rate is very low.

Other heat effects. The heat of crystallization can cause changes of 20 percent or more in the airflow and heat load of a dryer. It has the same value but opposite sign as the heat of solution. Substances with positive heats of solution give off heat when dissolved and take it back when dried, so this property affects results and should not be ignored. If values of the heat of solution are in kg · cal/g · mol (as in Perry and Green, 1984), they can be converted to heat of crystallization in Btu/lb by multiplying by −1800 and dividing by the molecular weight, including any hydrated water remaining in the product. Heat of solution values found in the literature are for infinite dilution; thus the full value is too safe when negative and unsafe when positive. Correct values can be obtained using a calorimeter.

In a similar way other heat losses or gains can be accounted for by converting them to Btu/lb of product, including its moisture. The heat of reaction is one example; another is the heat added by an internal heat exchanger. These effects can be totaled as Btu/lb of product and entered for heat of crystallization. A positive value adds heat to the system, which reduces the heat the air has to supply. This allows either reduced airflow or a lower inlet air temperature for the same production rate. A negative value has the opposite effect.

An example of the need to check all input details is the spray drying of ceramics in tower dryers. In addition to heat loss from the tall tower, the solids leave at a much higher temperature than the entering feed. And, as the drying air enters, its temperature is lowered by the compressed air used for atomization—a leak, in effect, at a critical point.

5.2.4 Calculation procedure

The calculation procedure considers the rates, properties, and temperatures for air and product and all essential operating conditions, including plant elevation, collector efficiencies, and an inward leak. To complete the procedure for a specific system, other items can be added, such as calculations for scaleup, equipment features, dust loadings, and conversion of units. Some of these options are described in Sec. 5.2.8.

Heat and material balance.   The heat and material balance equations are the core of the calculations. They permit solving for the outlet air moisture and airflow rate. The equations are developed from the following relation: heat entering the dryer in the air and feed, adjusted for heat loss or gain, equals heat leaving in the air and product. Any heat change that will have a significant influence is included. When those results are known, the flow rates, heat load, and saturation conditions at the dryer outlet can be calculated. Air enthalpy equations show only a single temperature, rather than a difference of two temperatures, because the reference temperature is 0.0°F.

Solution of the heat and mass balance requires trial-and-error loops, converging on outlet air enthalpy. Airflows and vessel surface area are needed but not known at the start, so they are estimated for the first loop. For this procedure the vessel is assumed to be a cylinder with both ends flat or one conical, but any geometry could be substituted. Ordinarily only two or three loops are needed to converge.

Enthalpies and specific heats.   The equations use the enthalpy of each material above a reference temperature—0°F for air and 32°F for water. Metric systems have a less complex, single basis of 0°C for both air and water.

The enthalpies of air and combustion gas diluted with air are similar, because nitrogen is dominant and properties are not much different. Saturation conditions differ to some extent, but air enthalpies and average molecular weight do not, except when combustion gases are used for drying with little or no dilution by air.

Various data are listed in App. B, including dry air enthalpies, which vary by more than 4 percent from 100 to 1000°F. To be consistent with the sensible heats of other materials in the equations, they are converted to specific heats. For specific heats of solids and water, it is satisfactory to use single-point values because they vary little for their temperature ranges in industrial dryers.

Bases of procedure

1. Definitions are given in the nomenclature in Sec. 5.2.2.

2. With a few exceptions, units are in the U.S. customary system. Appendix A gives conversions.

3. The temperature basis is 0°F for air and 32°F for water and solids.

4. The weight basis is 1.0 lb of dry air for enthalpies, air moisture, and humid volumes.

5. The feed and product moisture contents are on a wet basis.

6. Accuracy to three or four places is adequate for most design work and plant operations, and most data from tests are no better. But some intermediate calculations need greater accuracy, such as for wet-bulb and dew-point temperatures.

7. All percentages are assumed to be reduced to fractions before calculations begin.

8. No safety factor has been included, but one is recommended for heater, blower, feeder, and any other items to avoid bottlenecks or to satisfy other concerns.

9. Equations for flow rates of feed, evaporation, and dry solids assume that the product rate is fixed. If product is lost (such as dust from the collectors), the feed and other rates are increased as necessary to keep the specified product rate.

10. Equations for W_i, H_t, P_t (from E_l), and estimates for T_w, T_p, and F_5 are empirical, as are the equations for P_e and P_p (Keenan et al., 1978*).

---

*Equations for P_e and P_p [Eqs. (5.75) and (5.76)] were taken from the 1936 edition of this work, authored by Keenan and Keyes.

5.2.5 Equations for calculating air conditions

System pressure. If pressure is not input, it is assumed to be atmospheric and is corrected for elevation.

$$P_x = 1.0018e^N - \frac{3.6}{E_l + 2000} \qquad N = -3.73832 \times 10^{-5} E_l \qquad (5.7)$$

$$P_t = 14.696P_x \tag{5.8}$$

Solid and liquid flow rates

$$E_t = E_a + (1 - E_a)E_b + (1 - E_a)(1 - E_b)E_c \tag{5.9}$$

$$W_p = P_r P_m \tag{5.10}$$

$$S_p = P_r - W_p \tag{5.11}$$

$$D_s = \frac{S_p}{E_t} \tag{5.12}$$

$$F_d = \frac{D_s}{F_s} \tag{5.13}$$

$$P_l = \frac{D_s - S_p}{1 - P_m} \tag{5.14}$$

$$W_f = F_d - D_s \tag{5.15}$$

$$E_v = F_d - P_r - P_l \tag{5.16}$$

Gross product P_g includes loss from collectors. Net product P_n is the input value and is also the collector output.

$$P_g = \frac{D_s}{1 - P_m} \tag{5.17}$$

$$P_a = E_a P_g \tag{5.18}$$

$$P_b = E_b(P_g - P_a) \tag{5.19}$$

$$P_c = E_c(P_g - P_a - P_b) \tag{5.20}$$

$$P_n = P_a + P_b + P_c \tag{5.21}$$

Estimated product temperature. T_p is estimated from T_w, adjusted for the effect of P_m, and used as the final value.

$$T_w = 164 - \frac{16{,}900}{T_2} \tag{5.22}$$

$$T_q = T_5\left(\frac{134.4}{T_w} - 0.834\right) + 2.4T_w - 192 \tag{5.23}$$

$$T_p = T_q - \frac{(T_q - T_w)P_m}{0.06} \tag{5.24}$$

If $\qquad P_m > 0.06 \qquad$ then $T_p = T_w$ $\qquad$ (5.25)

Air specific heats, enthalpies, and moistures. Subscript i represents stations 1 through 5; j is the subscript for air enthalpies. For

$$j = \text{int}\left(\frac{T_i}{100}\right) + 1 \tag{5.26}$$

select the lower of the two elements in the air enthalpy array that straddle the value for T_i. For $T_i < 200°F$, $C_i = 0.2400$.

$$C_i = \frac{K_j + (K_{j+1} - K_j)[T_i - \text{int}(T_i/100)100]/100}{T_i} \tag{5.27}$$

$$W_i = 1061.8 + 0.433T_i + 0.000041T_i^2 \tag{5.28}$$

$$H_1 = C_1T_1 + M_1W_1 \tag{5.29}$$

$$H_4 = C_4T_4 + M_4W_4 \tag{5.30}$$

If heating is indirect, $M_f = 0$, skip to Eq. (5.33).

$$R_c = \frac{N_h}{2}\left(\frac{M_v}{M_h}\right) \tag{5.31}$$

$$M_f = \frac{R_c}{H_v}\left(\frac{T_2C_2 - H_1 + M_1W_2}{1 - R_cW_2/H_v}\right) \tag{5.32}$$

$$M_2 = M_1 + M_f \tag{5.33}$$

$$H_2 = C_2T_2 + M_2W_2 \tag{5.34}$$

If using a direct-fired heater, correct for the effect of M_1. Increase H_2 by

$$0.46M_1(T_2 - T_1) \tag{5.35}$$

and revise M_2 to

$$M_2 = \frac{H_2 - T_2 \, C_2}{W_2} \tag{5.36}$$

Then
$$M_f = M_2 - M_1 \tag{5.37}$$

Estimated vessel size. V_f' is for a cylindrical vessel with or without a cone. D_m will be recalculated later.

$$L_t = L_p + L_k \tag{5.38}$$

$$F_5 = \frac{1500 E_v (1 + L_t)}{T_2 - T_5} \tag{5.39}$$

$$V_f = 0.785 \left(H_d + \left| \frac{\tan R_a}{6} \right| \right) \tag{5.40}$$

$$D_e = \left(\frac{F_5 E_x}{60 V_f} \right)^{1/3} \tag{5.41}$$

$$D_m = D_e \tag{5.42}$$

Estimated airflow rates. These are needed for the first loop in the heat and material balance calculation. Alternatively, guess D_m, A_5, A_2, and A_4, then correct them after the heat and mass balance and rerun the calculation. Skip after the first loop.

$$A_5 = \frac{F_5}{18} \tag{5.43}$$

$$A_2 = \frac{A_5}{1 + L_t} \tag{5.44}$$

$$A_4 = A_2 L_t \tag{5.45}$$

Heat and mass balance loop—outlet air moisture. Iterate until H_5 does not change significantly.

Estimate heat loss and heat load; to be recalculated.

$$A_r = 0.785 D_m^2 (1 + \frac{1}{\cos R_a} + 4 H_d \tag{5.46}$$

$$H_t = \frac{T_5 + 570}{7042} \tag{5.47}$$

$$Q_l = H_t \, A_r (T_5 - T_1) \tag{5.48}$$

$$Q_t = 2000 E_v \tag{5.49}$$

Return here with revised D_m to recalculate station 5 moisture and enthalpy. Skip to Eq. (5.51) if H_l was an input.

$$H_l = \frac{Q_l(D_m/D_e)^2}{Q_t} \tag{5.50}$$

$$Z_h = \frac{60A_2H_2}{1 + H_l} + 60A_4H_4 \tag{5.51}$$

$$Z_b = S_pC_sT_p + W_p(T_p - 32) - D_sC_sT_f - W_f(T_f - 32) - D_sH_c \tag{5.52}$$

$$Z_m = \frac{M_2 + L_t M_4}{1 + L_t} \tag{5.53}$$

$$M_5 = \frac{(Z_h - Z_b)/(60A_5) - C_5T_5}{W_5} \tag{5.54}$$

$$M_3 = (1 + L_t)M_5 - L_t M_4 \tag{5.55}$$

$$H_5 = C_5T_5 + M_5W_5 \tag{5.56}$$

Airflows, station 3 conditions, and heat load

$$A_5 = \frac{E_v}{60(M_5 - Z_m)} \tag{5.57}$$

$$A_2 = \frac{A_5}{1 + L_t} \tag{5.58}$$

$$A_1 = A_2 \tag{5.59}$$

$$A_3 = A_2 \tag{5.60}$$

$$A_4 = A_2L_t \tag{5.61}$$

Station 3 (needed only if auxiliary air or leak),

$$T_3 = T_5 + L_t(T_5 - T_4) \tag{5.62}$$

$$M_3 = M_5 + L_t(M_5 - M_4) \tag{5.63}$$

$$H_3 = H_5 + L_t(H_5 - H_4) \tag{5.64}$$

$$T_3 = \frac{H_3 - M_3W_3}{C_3} \tag{5.65}$$

Could iterate for exact value, but this is close enough.

Volumetric airflows (for stations 1–5). Note that V_i is ft³/lb DA, not the reciprocal of density D_y.

$$V_i = \frac{10.73(T_i + 460)}{P_t}\left(\frac{1}{M_a} + \frac{M_i}{M_v}\right) \tag{5.66}$$

$$F_i = A_i V_i \tag{5.67}$$

$$D_y = \frac{1 + M_5}{V_5} \tag{5.68}$$

Dryer dimensions (for a spray dryer). Save old value of D_m.

$$D_e = D_m \tag{5.69}$$

$$V_u = \frac{E_x F_5}{60} \tag{5.70}$$

$$D_m = \left(\frac{V_u}{V_f}\right)^{1/3} \tag{5.71}$$

Compare current and previous H_5. End loop and go on to heat load if they agree; otherwise return to Eq. (5.50).

$$Q_t = 60A_2(H_2 - H_1) \tag{5.72}$$

Saturation moisture.   For any value of T_i,

$$T_k = 273.16 + \frac{T_i - 32}{1.8} \tag{5.73}$$

$$T_x = 647.27 - T_k \tag{5.74}$$

$$P_e = \frac{C_n T_x}{T_k}\left(\frac{C_a + C_b T_x + C_c T_x^{3}}{1 + C_d T_x}\right) \tag{5.75}$$

$$P_p = \frac{14.696 W_c}{e^{P_e}} \tag{5.76}$$

$$M_s = \frac{P_p M_v}{M_a(P_t - P_p)} \tag{5.77}$$

$$W_i = 1061.8 + 0.433T_i + 0.000041T_i^{2} \tag{5.78}$$

$$H_i = 0.24T_i + M_s W_i \tag{5.79}$$

Wet-bulb temperature.   Converge H_x on H_i by iterating from $T_i = 200$. Calculate M_s and W_i from Eqs. (5.73)–(5.78).

$$H_x = H_5 + M_s(T_i - 32)\left(1 - \frac{M_5}{M_s}\right)$$ (5.80)

When $H_x = H_i$ then $T_w = T_i$ (5.81)

Saturation temperature and adiabatic saturation ratio.   Converge H_i on H_5, iterating from $T_i = T_w$. Get M_s and H_i from Eqs. (5.28), (5.29), (5.63), (5.64), and (5.73)–(5.78).

When $H_i = H_5$ then $T_s = T_i$ (5.82)

$$A_s = \frac{M_5}{M_s}$$ (5.83)

Dew point and relative humidity.   Converge M_s on M_5 by iterating from $T_i = T_s$. Get P_p and M_s from Eqs. (5.28), (5.29), (5.63), (5.64), and (5.73)–(5.78).

When $M_s = M_5$ then $T_d = T_i$ (5.84)

Get P_p at T_5 from Eqs. (5.73)–(5.76).

$$\mathrm{RH} = \frac{100 M_5 P_t}{M_v(1/M_a + M_5/M_v)P_p}$$ (5.85)

5.2.6 Vessel size

In general, weight and volumetric airflow rates determine vessel size and heat load. Calculated airflow, however, is only a minimum that satisfies the psychrometric relations, and often has to be increased to satisfy other conditions. The effect of air moisture on the product and on possible condensation is also important, as well as, for some systems, the need to convey the product. To satisfy these strictures, changes may have to be made in the flow or temperature of the air or in system design, sometimes in all three.

In *flash dryers* the diameter is found from the required airflow and minimum velocity. The heat transfer rate, found by testing, governs the length of the drying duct, which, in turn, determines the exposure time. Sometimes the airflow has to be increased (and the inlet temperature lowered) to convey the solids better or to move the outlet condition away from saturation. The inlet air temperature is lowered enough to hold the heat supply constant.

Fluid-bed chambers are sized to set a minimum exposure time for drying as well as lower and upper limits of air velocity for fluidizing. If either forces an increase in airflow, the heat supply is kept from increasing by reducing the inlet temperature, as for flash dryers. The excess heat can also be offset, if the system has an internal heat exchanger, by decreasing its surface.

Rotary dryer design considers the peripheral velocity of the shell as found by testing, because it affects the showering of solids. The slope of the shell is set to give the same forward leaning angle for the flow of solids off the flights. Most evaporation occurs while hot air passes through the solids, so these conditions are especially important for heat transfer dominated drying. Solids loading in the shell and exposure time are other important factors. For any test dryer or commercial dryer of the bed type, the vessel loading volume, measured during shutdown, divided by the volumetric flow rate determines the exposure time.

In sizing a *spray dryer* the outlet air moisture, as measured by relative humidity or some other indicator, has to be kept below a maximum for virtually all products. Various moisture parameters are discussed in Sec. 5.2.8. To control humidity, the outlet air temperature may have to be adjusted, and this will change the airflow rate. When the airflow is set, the chamber geometry can be determined, but it must include enough space to dry the largest drops. The volume and diameter can be calculated from the airflow and exposure time using Eqs. (5.70) and (5.71).

Another method is to select from standard chambers one with enough volume for at least the minimum exposure time at the required airflow. Small changes in air outlet temperature can generally be made to allow some latitude in exposure time.

A typical spray dryer design has a cone with a 60° angle and a cylinder height 0.55 to 1.0 times the diameter. For this geometry an exposure time in seconds of about two-thirds the diameter in feet suits many products; few need less, but some need up to about 50 percent more.

Exposure time scaleup. The time the solids spend in the dryer is an important measure of drying, and it is a key to scaling up many large units. In bed-type dryers the volumetric airflow rate divided into the total volume of the drying vessel gives the exposure time. In direct dryers that air-convey the solids, it is not possible to measure the exposure time of the solids. Instead, a calculation has to be made of the time the air takes to flow through the vessel. The conveyed solids may lead or, more likely, lag the airflow.

For a commercial unit the exposure time may be arrived at by experience factors without scaleup calculations. Some designs can use, without change, the time found by testing to be adequate. Because drying continues as the product is conveyed to the collector, true exposure time is extended—an important factor when exposure in the dryer is short.

The question is often asked why a longer exposure time is needed to make the same product in a larger-diameter spray dryer. With higher feed and airflow rates, the initial mixing of air and spray is less efficient. Second, when small particles are desired, it is increasingly difficult to attain small droplet sizes as chambers get larger.

Scaleup by heat transfer rate.   For heat transfer dominated drying, the exposure time is less significant, although usually still a factor. The design of rotary dryers and some other types is based on heat flow, which is used to calculate the vessel length. A common method is to consider the dryer as a heat exchanger, and to solve for the required area, similar to the indirect dryer method discussed in Sec. 5.1. But only the net drying heat flow is used. The net flow is the total heat load minus heat in the exhaust and heat losses, which are high if a large portion of the shell is uninsulated.

The cross-sectional area is the curtain of powder showered off the flights and is a function of the flight design. The area required is found by dividing the net heat load by a heat transfer factor. The latter is heat transfer rate times temperature difference, a combined factor either found from testing or based on experience with similar materials. The length is then determined from the area required and the flights that give the desired showering characteristics.

5.2.7   Example calculations

The input values in Table 5.3 and the results presented in Table 5.4 are for an example worked by computer. All 28 possible inputs are

TABLE 5.3   Input Values for Sample Calculation

E_l	3000 ft	N_h	4.30
E_a	70%	M_h	18.14
E_b	80%	H_v	23,651 Btu/lb
E_c	90%	L_p	0.0%
P_r	4000 lb/h	L_k	5.0%
P_m	2%	H_d	1.0
F_s	50%	R_a	60.0°
T_2	800°F	E_x	20.0 s
T_5	200°F	H_l	0.0%
T_1	80°F	C_s	0.4 Btu/(lb · °F)
T_4	60°F	T_p	0.0°F
C_p	0.0 Btu/(lb · °F)	T_f	120.0°F
M_1	0.007 lb/lb DA	H_c	−200.0 Btu/lb product
M_4	0.01 lb/lb DA	P_t	0.0 lb/in^2

TABLE 5.4 Example Dryer Results (U.S. Customary Units)

Miscellaneous data

Feed temperature	120.0°F
Product temperature	156.9°F
Solid specific heat	0.400 Btu/(lb · °F)
Elevation	3000 ft
Elevation correction factor	0.895
Feed solids	50.0%
Product moisture	2.0%
Heat of crystallization	-200.00 Btu/lb
Heat loss	0.28%
Calculated diameter	16.2 ft
Volume	4277.5 ft^3
Volumetric factor	1.012
Cylinder height-to-diameter ratio	1.00
Cone angle	60.0°
Exposure	20.0 s
Auxiliary air	0.0%
Air leak	5.0%
Rates, lb/h	
Feed	7887.3
Product	4000.0
Solids	3943.7
Evaporation	3863.2
Product loss	24.1

Psychrometric data at 13.150 lb/in^2 (absolute)

Dry gas molecular weight	28.97
Specific heat at station 2	0.2459 Btu/(lb · °F)

Station	Temperature, °F	Moisture, lb/lb DA	Enthalpy, Btu/lb DA	Humid volume, ft^3/lb DA	Dry airflow, lb/min	ACFM, ft^3/min
1	80.0	0.01000	30.13	15.45	535.0	8268
2	800.0	0.02877	238.00	37.13	535.0	19865
3	205.5	0.14912	221.21	23.24	535.0	12433
4	60.0	0.00700	21.98	14.81	26.8	396
5	200.0	0.14235	211.73	22.84	561.8	12833

Collector	Product, lb/h	Efficiency, %
1	2817	70.0
2	966	80.0
3	217	90.0
Net	4000	99.4

Heat and outlet air moisture data

Number of hydrogen atoms	4.30
Molecular weight	18.14
Higher heating value	23,651
Fuel moisture	0.0188 lb H_2O/lb DA
Saturation moisture	0.1594 lb H_2O/lb DA
Heat load	6.6728 MBtu/h

Dryer outlet (station 5)

Adiabatic saturation ratio	89.3%
Relative humidity	21.2%
Dew-point temperature	133.7°F
Saturation temperature	137.1°F
Wet-bulb temperature	137.5°F

listed. They are given in the order they occur in the equations. This is convenient for manual calculations. For computer input it is better to arrange items by categories on a form, which helps to eliminate errors. Input values were chosen to illustrate their use, rather than to be typical. Collector efficiencies E_a, E_b, and E_c, for instance, were set low to show their cumulative effect, the product distribution, and the loss. Zero entries for C_p, H_l, T_p, and P_t signify that each will be calculated by the program.

Example results. Table 5.4 tabulates the results and most of the inputs for the example. Included is the selection of a spray-drying chamber having an exposure time of 20 s.

When results of a calculation are not as expected, it is well to recheck the input values. If no errors are found in the input, the results should be checked carefully—they may indeed be correct. Third, and most difficult, the procedure should be reviewed, especially if it has been converted to a computer program. Programs do not display each step, as in a manual calculation, and complex programs cannot be tested in every possible path; thus errors are found even after long use.

Having moisture, enthalpy, and humid volume on the basis of a pound of dry air simplifies the equations and calculations. But for some purposes humid volume has to be converted to specific volume (reciprocal of density), which is cubic feet of the mixture of dry air and its moisture per pound of the mixture. In the example, the density at station 5 is 0.05002 lb/ft^3 from Eq. (5.68); thus the specific volume is 19.99 ft^3/lb.

Feed at 120°F contributes some heat to the system. This reduces both the flow of air and the heat it supplies. Because the heat loss was entered as zero, it defaults to a calculation by an equation assuming a 2.0-in-thick standard insulation. The heat loss is only 0.28 percent, which has little effect on the results. On the other hand, the negative heat of crystallization increases both the airflow and the heat flow substantially. Elevation at 3000 ft increases the volumetric airflow rate by nearly 12 percent compared to operation at sea level.

Entry of a 5 percent leak at 80°F air at station 4 ends the drying (theoretically) at 205.5°F at station 3, which is calculated back from station 5 conditions and the leak.

5.2.8 Other calculations

The equations can be used—in a modified form—to find data points for psychrometric charts. Furthermore to generate data for shortcut charts, the procedure can be used as given, making practical assumptions for heat effects such as feed and product temperatures.

Multiple calculations are simplified if the procedure is converted to a computer program. For reruns it can be arranged for the entry of only the items that are to be changed. Studies can be made of a particular product, or the effect of one set of variables on another, and in this way much can be learned in a short time. (See the results of the dryer humidity study in Table 5.5.)

To complete a program for a specific design, other routines have to be added; some are described hereafter. In addition, some of the data listed in App. B can be added. Longer programs can be made more general if the option is provided to enter data and print results in U.S. customary units and in SI or other metric units.

Recycling of air or product, and even more elaborate schemes, could be added to the procedure. Evaporation of liquids other than water, however, requires extensive changes, the most important of which are given in Chap. 13. Other possible applications include finding the best conditions during testing and evaluating various dryer types to optimize investment and operating costs. The latter are discussed in Chap. 8.

Calculating leaks and auxiliary airflows. Unintended leaks of air into a dryer are hard to determine, although operational upsets sometimes give clues. As an example, if the capacity is too low, there may be a leak between the temperature-sensing point and the dryer inlet. Such leaks drop the inlet temperature below the reading. Or there may be a leak into a cyclone collector that is reducing its efficiency and causing it to lose product.

Sometimes an inflow can be located and gauged by calculating the difference in the airflow rates before and after the suspected location. For calculating, known leaks that enter the drying chamber should be combined with the dryer air.

One way to calculate the airflow at the dryer inlet involves first accurately measuring T_2 and the flow rate of the heat source. Using a psychrometric chart, M_1 is found from measured values of T_1 and T_w. M_2 is calculated, and the air enthalpies at stations 1 and 2 are calculated or read from the chart. Then the heat load and airflow at the dryer inlet are calculated (and confirmed by measurement if practical). Using the shortcut equation, Eq. (5.86), for combustion moisture (see App. B for S_l),

$$M_f = (T_2 - T_1)S_l \qquad (5.86)$$

(Assume M_1 does not affect M_f.)

$$M_2 = M_1 + M_f \tag{5.87}$$

$$Q_t = \text{heat source rate times its heating value} \tag{5.88}$$

$$A_2 = \frac{Q_t}{60(H_2 - H_1)} \tag{5.89}$$

If steam is the heat source, its flow rate must be adjusted for flashing of vapor. The airflow at the dryer outlet can be determined by the pressure drop in a collector and confirmed by a fan curve, or it can be computed from velocity measurements; but they require finding the humid volume from a psychrometric chart. It can also be calculated from the data used in the preceding plus the evaporation rate and the wet-bulb temperature at the dryer outlet. (See Chap. 7 for measuring wet-bulb temperatures.) The intersection of T_w and T_5 locates M_5 on the chart. Then the airflow at the dryer outlet is

$$A_5 = \frac{E_v}{60(M_5 - M_2)} \tag{5.90}$$

Dryer outlet humidity. An important influence on product quality is the outlet air humidity. Although seldom measured in plant operations, it is considered to be acceptable as long as the product meets its specifications. The testing of products, in most cases, sets the operating conditions, and tests are generally run at only one set of weather conditions. But a commercial operation has to make the same product at various outlet moistures, because the moisture in the supply air changes with the seasons. To keep the product the same, the dryer operating temperatures should be adjusted.

When the dryer outlet temperature is decreased, for example, experienced operators know that to keep the same product quality, the inlet temperature also has to be decreased. How much decrease is needed to attain the same quality, and which of the various measures is the best indicator of it, is not known. The amount of change needed differs with the nature of the product and can only be determined by testing. But some idea of the best way to measure the needed change can be found by calculations.

There are several measures, or indicators, of the dryer outlet condition, and the best known are dew-point and wet-bulb temperatures and relative humidity. (Absolute humidity and dew-point temperature are essentially the same measure.) A lesser known parameter is percent adiabatic saturation (ASR), which is absolute humidity over saturated moisture at the same enthalpy times 100.

Table 5.5 gives the results of one set of calculations that show how these measures are affected by drying temperatures. The basis was a

TABLE 5.5 Humidity Comparison

Temperature, °F			Absolute humidity, lb/lb DA	Dew-point tempera-ture, °F	ASR, %	Relative humidity, %
Inlet	Outlet	Wet bulb				
400	200	117.3	0.05299	106.6	72.8	10.0
400	190	117.3	0.05530	107.9	75.9	12.8
389	190	116.6	0.05292	106.6	74.9	12.3
368	190	114.3	0.04837	103.7	72.8	11.4
339	190	111.5	0.04209	99.4	69.4	10.0
Inlet change, °F. for −10°F outlet change				−11	−32	−61
Outlet change, °F, for +20°F inlet change				+18	+6	+3

dryer air temperature of 400°F inlet and 200°F outlet. The outlet was lowered to 190°F. The inlet temperature was then lowered incrementally to find the temperature at which each of the measures was returned to its original value.

The first line of the table is the basis. The second line shows the result of dropping the outlet temperature from 200°F to 190°F, with the inlet held at 400°F. All of the measures change except the wet-bulb temperature, and so it offers no clues on preserving the product. The third line shows that lowering the inlet temperature to 389°F returns the dew-point temperature to its original value.

Some experienced dryer engineers consider the adiabatic saturation ratio to be the most reliable measure for keeping the same product. To satisfy it in this example, the inlet temperature had to be dropped to 368°F. Relative humidity is widely used as a comparative measure, especially in dairy products. But it required what seems to be an unreasonably large change of −61°F to 339°F.

The final two lines in Table 5.5 summarize the effects of adjusting temperature to maintain a constant product. The next to the last line is for the outlet temperature, returning the dew point, ASR, and relative humidity to their original values by changing the inlet. The last line shows the results of a reversed set of calculations—changing the outlet temperature to correct a change in the inlet temperature.

Perhaps a modification of one measure or some combination of several would suit a majority of materials. To find the best answer for a single product or related products would require a series of tests.

Dust loadings. Dust loads can be calculated from the loss from each collector, with results converted to any desired set of units. To solve for dust loadings into and out of each collector requires the gross product including losses, P_g, and the products from each of the collectors, P_a, P_b, and P_c.

C_g is a lumped term (of inconsistent units) to convert lb/h to gr/ft^3. G_a, G_b, and G_c are dust loads into the three collectors, and G_d is the dust load out of the third. The equations are

$$C_g = \frac{7000}{60F_5} \qquad (5.91)$$

$$G_a = P_g C_g \qquad (5.92)$$

$$G_b = (P_g - P_a)C_g \qquad (5.93)$$

$$G_c = (P_g - P_a - P_b)C_g \qquad (5.94)$$

$$G_d = (P_g - P_a - P_b - P_c)C_g \qquad (5.95)$$

Scrubber evaporation. Wet scrubbers can perform several functions as well as capturing fines. One is to evaporate water from feed passed through it on the way to the dryer. This relieves the dryer of that much duty; thus it reduces dryer size and operating cost. But the feed has to be below its maximum concentration and must not foam or build up on scrubber walls.

Evaporation in a scrubber E_s can be found using a psychrometric chart or equations. It is a function of the dryer outlet airflow rate A_5 and the difference between air moisture M_5 at the dryer outlet and saturation moisture M_s, at the scrubber outlet enthalpy. The equation is

$$E_s = A_5[60(M_s - M_5)] \qquad (5.96)$$

This relation is slightly conservative because, theoretically, M_s should be figured at the wet-bulb temperature. That would yield slightly more evaporation, but the benefit is offset if the scrubber is less than 100 percent efficient and does not saturate the air.

When spray to the scrubber is sufficient to cool the airstream beyond the point where saturation is reached, condensation begins. This can continue until the unit becomes a net condenser rather than an evaporator. This principle is used in self-inertizing and other recycle systems to condense moisture. The saturated air from a well-designed unit can be within 5 to 10°F of the incoming cooling water.

Mixing airstreams. The properties of a mixture of two or more airstreams can be found if the temperatures, absolute moistures, and flow rates by weight are known for each. Volume flow can be converted to weight flow using

$$A_i = \frac{F_i}{V_i} \qquad (5.97)$$

The method for finding the mixture is as follows.

1. Solve for the enthalpy of each stream, using Eqs. (5.78) and (5.79) or a psychrometric chart.

2. Multiply each moisture by its fraction of the flow rate. The sum of these is the moisture of the mixture.

3. Do the same for the enthalpies to get the enthalpy of the mixture.

4. The temperature of the mixture can be estimated in the same way. It will be approximately correct if the temperatures of the components are not far apart. A more accurate value can be obtained by writing a heat and mass balance, equating the two flows to the mixture. It can also be interpolated from a psychrometric chart.

Combustion moisture. Moisture added to a dryer by a direct-fired heater can be found by first computing R_c, the weight ratio of moisture formed to fuel,

$$R_c = \frac{N_h M_v}{2 M_h} \qquad (5.98)$$

where N_h is the number of hydrogen atoms in the chemical formula of the fuel, and M_v and M_h are the molecular weights of water vapor and fuel. The equation for M_f, moisture formed by combustion, was given in Sec. 5.2.5,

$$M_f = \frac{R_c}{H_v} \left(\frac{T_2 C_2 - H_1 + M_1 W_2}{1 - R_c W_2 / H_v} \right) \qquad (5.32)$$

In this equation H_v is the higher heating value of the fuel in Btu/lb fuel, T_2 and C_2 are the temperature and specific heat of the dry air, and W_2 is the water vapor enthalpy (from steam tables), all at station 2. M_1 is the air moisture at station 1, and H_1 is the air enthalpy from Eq. (5.29).

Especially in cases of high ambient humidity, this value should be corrected for the influence of the supply air's moisture M_1 on the fuel required, thus on the moisture formed. See Eqs. (5.35) to (5.37). Then

$$M_f = M_2 - M_1 \qquad (5.99)$$

Percent oxygen in air from a direct-fired heater. From the chemical equation for combustion it can be shown that the stoichiometric ratio of weight of water formed per weight of inert M_x is

$$M_x = \frac{18.02N_h/2}{44.01N_c + 3.7733(N_c + N_h/4)28.163} \qquad (5.100)$$

where N_c and N_h are the number of carbon and hydrogen atoms, respectively, in the chemical formula of the fuel. For fuel mixtures they have to be averaged. The average molecular weight of nitrogen and other inerts in air is 28.163, and their mole and volume ratio to oxygen is 79.05/20.95, or 3.7733.

When excess air is added, the combustion moisture in the total air can be found from Eqs. (5.32) and (5.99) for M_f. M_x and M_f are ratios of the same moisture for stoichiometric air and total air. Thus M_x/M_f is the ratio of total air to stoichiometric air, and M_x/M_f - 1 is the fraction of excess air to stoichiometric air. Then, with 23.15 as the weight percent of oxygen in normal air,

$$\text{wt. \% of oxygen} = \frac{M_x/M_f - 1}{M_x/M_f} \times 23.15 \qquad (5.101)$$

Equation (5.101) assumes that the amount of air is unchanged by combustion. However, the air out of the heater contains the same nitrogen as that entered, but less oxygen and added carbon dioxide. The final result depends on the fuel. For the accuracy normally required, it can be considered that the fuel converts to water vapor and the air is unchanged.

6

Psychrometric Charts

6.1 Introduction

Psychrometric charts translate data into a convenient graphic form, and they clarify those drying systems that use a gas to supply some or all of the heat. They are also useful for heating, cooling, condensing, and some other process steps. These charts show the property relations of a gas (most often air) and a vapor (most often water vapor) up to their saturation conditions. Although their use for final dryer calculations has declined with the advent of computers, they help to make estimates and check calculations. Without them it would be hard to visualize and confirm some drying operations.

This chapter describes methods for both illustrating and calculating drying operations for air and water vapor on psychrometric charts (in U.S. customary units). Also shown are the operations of wet scrubbers, condensers, and internal and external heat exchangers, and the mixing of two airstreams. In addition, shortcut charts are introduced, constructed specially to allow a direct estimation of a dryer's airflow and heat load. In Chap. 13 standard and shortcut charts for nonaqueous solvents are described.

A chart point can be located if any two values of wet- or dry-bulb temperature, moisture, enthalpy, or humid volume are known. The latter three are on a basis of 1 lb of dry air (DA). Moisture is given in lb water vapor/lb DA, and enthalpy is the total heat in the air and its vapor in Btu/lb DA. Humid volume is the total volume of air and its vapor in ft^3/lb DA. It can be converted to density by inverting and multiplying by $(1 + M)$, where M is the moisture in the air in lb/lb DA.

Although useful in many ways, each chart can represent only one solvent vapor in one gas at one pressure, and accuracy is limited by the scale of the plot, especially when interpolation is required. In addition, the techniques needed for calculating various heat effects are

time-consuming to perform; but without them the charts are restricted to the simplest drying problems.

Types of psychrometric charts.   Several forms of psychrometric charts
can be found in drying practice. All use the same data, but are plotted
differently. *Mollier charts* plot moisture against enthalpy on oblique-
angle coordinates. These are in metric units and enjoy widespread use
in Europe and elsewhere outside the United States. (A point of confu-
sion is that in the United States Mollier diagram refers to an enthalpy-
entropy chart.)

The most common psychrometric chart in the United States has several variations of plots on regular coordinates of temperature in °F (x
axis) against moisture in lb/lb DA (y axis). Lines of constant enthalpy
in Btu/lb DA have a negative slope. These charts in the low temperature ranges are used widely for comfort heating and cooling applications, and for drying—usually at relatively low temperatures.

Figures 6.1–6.4 plot enthalpy against moisture and are more useful for drying, especially at higher temperatures. They are on regular coordinates, and lines of constant temperature have a positive
slope. All the lines of constant enthalpy, moisture, wet- and dry-
bulb temperature, and humid volume can be shown and all are
straight lines. With these advantages, they are the simplest to con-
struct and use when illustrating various heat effects, and they are
the easiest to understand.

All the air-water plots in U.S. customary units base enthalpies on
air at 0°F and water at 32°F. A basis of 0°F for air simplifies calculating the enthalpy. Using 32°F for water avoids the skewing of properties when water changes to ice. Metric systems are consistent in using
0°C as the reference for both air and water. The difference in bases
causes the metric system enthalpies to be 7.68 Btu/lb (17.86 kJ/kg)
less. The charts can be shown in only one system of units. Thus it is
not practical to give conversions throughout this chapter. (See App. A
for conversion constants.)

Drawing psychrometric charts.   Data for plotting air–water vapor psychrometric charts can be obtained using the equations given in Chap.
5. They have to be rearranged and solved for each variable, using multiple steps of 10 or 100 (or other appropriate values). If greater accuracy is needed, as for dryer design, the relevant portion can be plotted
on a larger scale. Property data are given in App. B. The calculation
method is detailed here.

1. Use selected stepped values of saturation temperature to calculate air specific heat, water vapor enthalpy, and saturated moisture
M_s. Equations (5.26)–(5.28) and (5.73)–(5.77) apply.

2. Calculate the enthalpy at saturation H_s for stepped values of saturation temperature, inserting data from step 1 into Eqs. (5.78) and (5.79). Using the values of H_s and M_s, plot the saturation points.

3. For dry-bulb temperature lines, calculate the enthalpy with Eqs. (5.26)–(5.29), using stepped values for temperature and moisture. Plot the temperatures below the saturation line.

4. Calculate the zero-moisture enthalpy H_z for wet-bulb temperature lines using the equation

$$H_z = H_s - (T_w - 32)\,M_s C_l \tag{6.1}$$

where the specific heat of liquid water C_l equals 1.0. Form the wet-bulb temperature line by connecting point H_z with its saturation point. Only a few wet-bulb lines are needed, but many temperature points should be marked on the saturation line to aid interpolation. At these points all temperatures meet—dry bulb, wet bulb, saturation, and dew point.

5. For humid volume rearrange Eq. (5.66) to solve for temperature. Calculate the values of temperature for stepped values of humid volume and moisture, calculate enthalpy, and plot the humid volume lines.

Note that lines of constant wet-bulb temperature are not parallel to lines of constant enthalpy, and the slope increases as the temperature rises. They differ at zero moisture by the enthalpy of the water evaporated (Treybal, 1980). The difference is equal to the enthalpy at saturation minus H_z from Eq. (6.1). Thus wet-bulb lines should not be considered as adiabatic drying lines. The difference is small below 90°F.

Gauging humidity.   Percent relative humidity is displayed on many psychrometric charts as a measure of distance from saturation. This is useful for comfort level, but is less satisfactory for drying operations, because it has very little meaning at high temperatures. A more useful measure is the adiabatic saturation ratio (ASR), which is the absolute humidity over saturation humidity at the same enthalpy. It is normally expressed as a percent, and 80 percent ASR at a given point, for example, is 80 percent of the distance between zero moisture and saturation on its enthalpy line. Divisions between equal ASR lines are evenly spaced, making the concept easy to visualize. Thus the lines can be omitted from these enthalpy-moisture plots, avoiding some clutter.

6.2   Indirect and Direct Heating

Figures 6.1–6.5 were prepared using the method outlined. In Fig. 6.1 two drying operations are shown, both at the same moderate drying

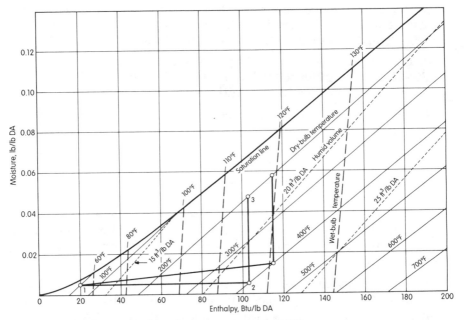

Figure 6.1 Psychrometric chart for temperatures to 600°F.

temperatures, but one heats the air indirectly, the other by direct firing with natural gas. Heating adds enthalpy, changing conditions
from station 1 to station 2. If the heating is direct-fired, it increases
the humidity in the air. This is because the combustion forms water
vapor, which can be calculated by Eqs. (5.31)–(5.37). Drying adds
moisture—usually with little change in heat—moving the conditions
from station 2 to station 3.

Data for the two heating modes are listed on Table 6.1. Columns I
and II are from chart readings (the last digit is a rough estimate), with
no allowance for heat effects other than from the air and its vapor. Results for both are about the same, except that direct firing raises the
heat load about 12 percent and the dew point about 4°F. Column III
lists computer-generated results, allowing for typical feed and product
temperatures and radiant-convective heat loss. The chart and computer results compare closely, because the heat effects are relatively
minor—feed at 60°F, product at 153°F, and heat loss of 0.67 percent.
For more extreme effects the results could have been much different.

6.3   Elevation and Operating Pressure

Most continuous industrial dryers operate at the atmospheric pressure
where they are located. Ordinarily this is changed little by blowers

TABLE 6.1 Comparison of Indirect and Direct Heating*

Station	Temperature, °F	I Indirect heat (from chart)			II Direct heat (from chart)			III Direct heat (computer-calculated)			Equation
		M	H	V	M	H	V	M	H	V	
1	60	0.0050	19.8	13.2	0.0050	19.8	13.2	0.00500	19.80	13.21	(5.57)
2	400	0.0050	102.8	21.9	0.0143	114.3	22.2	0.01431	114.33	22.17	(5.67)
3	200	0.0476	102.8	17.9	0.0577	114.3	18.2	0.05664	113.16	18.15	(5.72)
Airflow, lb/min		391			385			392			(5.84)
Airflow, ft³/min†		7005			6989			7108			
Heat load, Btu/h		1,947,000			2,181,000			2,220,000			
T_{dew}†		103.2			109.3			108.8			

*Basis—1000 lb/h evaporation, 14.696 lb/in². Columns I and II—no heat loss or gain; column III—typical conditions: feed at 60°F, product at 153°F, heat loss 0.67%.

†Volumetric air flow and dew point are at station 3, dryer outlet.

89

and pressure drop in the system. Plant elevation may have a significant effect on pressure, however, and thus on humid volume and volumetric airflow [see Eq. (5.66)]. Elevation correction factors for pressure are given in App. B. At an elevation of 5000 ft, as an example, the factor is 0.832. At this elevation, compared to sea level, volumetric airflow is increased by 1/0.832, or 20.2 percent, for the same mass flow.

Pressure also influences saturation conditions. At the temperatures in most industrial dryers, compared to sea level, operation at 5000 ft reduces both dew-point and wet-bulb temperatures by 6–7°F. The reductions are about the same for the next 5000-ft rise. On the other hand, pressure has virtually no effect on enthalpy-moisture relations.

6.4   Recovering Heat from the Exhaust

Figure 6.2 compares two dryers, both heated by direct natural gas firing. Their conditions are the same, except that one system has an external heat exchanger that preheats the supply air with the dryer exhaust, which is cooled to just above its dew point. More heat could be reclaimed if some water were condensed, but for many applications this causes problems, such as fouling and corrosion.

Drying without an exchanger is shown by the lines joining stations

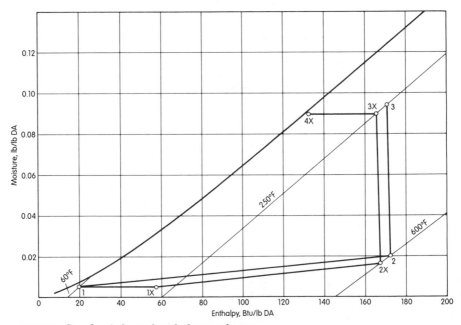

Figure 6.2 Supply air heated with dryer exhaust.

1, 2, and 3, which identify supply air, dryer inlet, and dryer exhaust. Stations 1, 1X, 2X, and 3X show the same dryer with the exchanger, which cools the exhaust to station 4X. The exchanger heat load is the enthalpy difference between stations 3X and 4X (just shy of the dew point). This enthalpy difference is added to the supply air, thus locating station 1X and the temperature of the air leaving the exchanger on its way to the heater.

For direct-fired heating, as in this example, the heat saving is greater than the exchanger's heat load. The extra benefit is the difference in enthalpy between stations 2 and 2X, and is the result of burning less fuel, thus forming less moisture. The added saving is about 12 percent of the exchanger duty for natural gas firing and 6 percent for fuel oil.

6.5 Influence of Various Heat Effects

A vertical line on these charts represents adiabatic drying, with simultaneous cooling of the air and no heat loss or gain. Drying lines become angled to the left of vertical by heat losses, and to the right by heat gains. Neither of the changes on the previous two charts—heat source and external exchanger—affected the angles of the drying lines. Thus moisture levels in the air stayed the same, and only the heat load changed. But in Fig. 6.3 heat effects do alter moisture levels, changing both heat load and mass airflow.

Three dryers are shown in Fig. 6.3 under the influence of three heat effects—surface heat loss, heated feed, and an internal heat exchanger. For all three examples heating is indirect, air temperatures are 300°F at inlet and 200°F at outlet, feed moisture is 50 percent, and product temperature is 150°F. The intersection of each drying line with the 200°F line is its dryer outlet point. When the moisture difference between stations 2 and 3 is increased, the extra moisture in each pound of air reduces both the airflow and the heat required. But a penalty is paid in higher humidity, which may affect product moisture.

Heat loss. Drying line *A* results from a 0.5 percent heat loss, about the expected value for an average sized, properly insulated dryer with 200°F outlet. The effect can be found graphically by reducing the station 2 enthalpy by the percent heat loss, then proceeding vertically from there to the outlet temperature line. Heat loss is seldom a good substitute for safety factor; too often it proves excessive. The 5 percent heat loss shown by the dashed line angled to the left increases both airflow and heat load by more than 15 percent.

If a high value of heat loss is chosen for a dryer with low air tem-

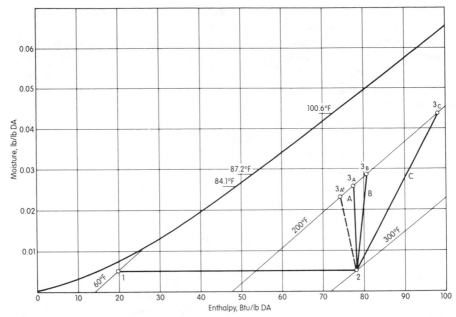

Figure 6.3 Various heat effects. *A*—heat loss only; *B*—heat loss and heated feed; *C*—heat loss and heat exchanger.

perature difference, it can cause a discontinuity in the calculation, and the moisture difference can become negative. Introducing excessive amounts of cold air to a dryer can cause the same problem. An important benefit of using psychrometric charts is the ability to see such events and possibly the corrections needed.

Other heat effects. Line *B* in Fig. 6.3 is for a dryer with the same 0.5 percent heat loss, but with the feed temperature raised from 60 to 160°F. To determine this graphically, both liquid and solid in the feed have to be converted to the equivalent heat in the feed per pound of moisture. The net value in Btu/lb moisture is the slope of the drying line drawn from station 2.

Drying line *C* represents a dryer with 26 percent of its total heat requirement provided by an internal heat exchanger, as in a fluid-bed dryer. Both heat and airflow requirements are greatly reduced, but the dryer outlet humidity is much higher.

These and other heat effects—some positive, some negative—can be combined into one value of heat per pound of water evaporated, and plotted on a chart such as Fig. 6.3. They include heat in the feed's solids and liquid, heat of crystallization or reaction, and heat from an internal exchanger. Heat loss must be handled separately.

Again, stations 1 and 2 are the supply air and dryer inlet. Station 3, the dryer outlet for each condition, is at the intersection of each drying line with the outlet temperature line. The method assumes that the product moisture content is small.

1. Select a ratio S_a of lb solid/lb DA to fit the moisture range of the chart. A value between 0.01 and 0.04 fits most conditions.

2. Multiply S_a times the ratio of water to solid in the feed to obtain W_a, the ratio of lb water/lb DA. For low feed moistures higher values of S_a will give greater accuracy.

3. Calculate W_e, the effect of the feed-water temperature in Btu/lb DA,

$$W_e = (T_f - 32)C_w W_a \tag{6.2}$$

where T_f is the feed temperature and C_w is the liquid-water specific heat in Btu/(lb · °F).

4. Calculate S_e, the effect of the feed-solids temperature in Btu/lb DA,

$$S_e = (T_f - T_p)C_s S_a \tag{6.3}$$

where T_p is the product temperature and C_s is the solid specific heat in Btu/(lb · °F).

5. Convert the heat of crystallization to Btu/lb solid. Values in the literature are usually given as heat of solution in kg · cal/g · mol and can be converted to H_c in Btu/lb solid by multiplying by (− 1800/ molecular weight). H_c can also be used for the heat of reaction if converted to Btu/lb solid. Then calculate C_e, the effect of this heat in Btu/lb DA,

$$C_e = H_c S_a \tag{6.4}$$

6. Heat from an internal heat exchanger can also be converted to Btu/lb solid, then to Btu/lb DA by multiplying by S_a. The heat it adds, Q_x, may be a known fraction of the total heat load. It can be figured by the ratio from that fraction times the enthalpy difference between stations 2 and 1. Then

$$X_e = Q_x S_a \tag{6.5}$$

Or, if the limiting dryer outlet humidity is known, this and the outlet air temperature identify station 3. If there are other heat effects, an iteration, using all the effects, will be needed to pinpoint station 3. The Btu/lb DA for the exchanger can be found from the chart as the difference in enthalpy between stations 3 and 2.

7. Algebraically add W_e, S_e, C_e, and X_e for the net effect N_e in Btu/lb DA.

8. Plot a point from station 2, first subtracting the enthalpy for any heat loss, moving upward a distance W_a, then N_e to the right if positive and left if negative. Station 3 is where the line drawn through station 2 and this point intersects the outlet temperature line.

6.6 Recycle

Figure 6.4 traces the path of a dryer that recycles a large portion of its exhaust. The main functions of the system are outlined in Fig. 2.4. Because of the added operations, station numbers are different from those used previously. After drying between stations 3 and 4, the air enters the scrubber condenser and leaves it at station 5. Bleed is removed from the airstream, and the remainder is mixed with enough supply air from the heater to bring the mixture to the dryer inlet temperature. The mass flow rate of bleed equals the supply air, plus any other inflows such as purge air to a bag filter collector. Total recycle, or closed-loop, operation is covered in Chap. 13.

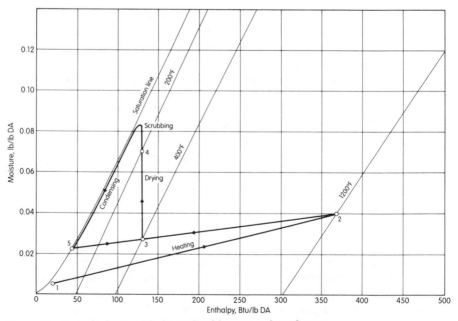

Figure 6.4 Recycle dryer with direct-fired heater and condenser.

6.7   Mixing Two Airstreams

The result of mixing two streams of gas at different conditions can be determined faster by chart than by calculating with a heat balance. For many jobs the accuracy is adequate. Mixing is seen in Fig. 6.4 by the straight line through stations 5, 3, and 2. The lengths of the line segments (or, more accurately, the enthalpy differences) are proportional to the two airflow rates. Note that increasing amounts of hot air move the mixture point farther from the cold end, and the amounts of hot and cold air are inversely proportional to the enthalpies, or line lengths.

A more complex situation is a large inflow of air into a dryer. This might be a large leak, or an addition of cold air to cool the product as drying ends. The procedure to be applied can be visualized using Fig. 6.2, for which the original station numbers apply. The station 1–2 line is heating; the station 2–3 line is drying without cold air, but considering any other heat effects. On the station 1–2 line, if 60°F cold air is added, for example, and is 20 percent of the hot drying air, the mixture will be 0.20 of the distance from station 2. Cold air to a dryer is figured as follows.

1. Locate station 2, dryer inlet, and the station for cold air (which may or may not be the same as station 1, heater supply air). Draw a line between the two stations, and on it compute point A, the location of the mixture of hot and cold air.

2. From point A draw a line parallel to the station 2–3 line. Dryer outlet, station 5, is where this new line intersects with the outlet temperature line.

3. Draw a line through stations 1 and 5 and extend it to meet the station 2–3 line. The intersection is station 3, which is at a higher temperature than the dryer outlet because cold air has entered.

6.8   High-Temperature Chart

Figure 6.5 is an air–water vapor psychrometric chart on log-log coordinates, chosen as the best way to reach 2500°F. Air temperatures to 2000°F and above are used on pulse jet dryers and calciners. A log-log plot allows a wide range of conditions, although with less accuracy at the high end. Gas composition also affects the accuracy slightly. It deviates from that of air because of the presence of combustion gases, but nitrogen dominates both gases, so average properties differ little.

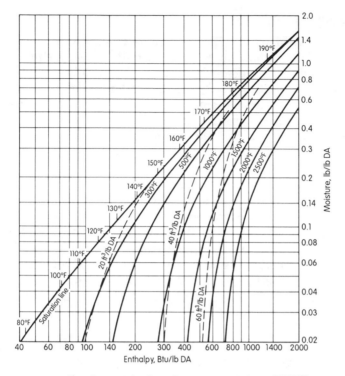

Figure 6.5 Psychrometric chart for temperatures to 2500°F.

6.9 Shortcut Charts

When air inlet and outlet temperatures are known, quick dryer esti-mates can be made with the nomographs in Figs. 6.6 and 6.7. Their curves were plotted from calculated values for typical drying condi-tions. Heavier lines are the dryer inlet temperatures, lighter lines the outlet temperatures. The charts yield volumetric airflow at dryer out-lets and heat load, the two essentials for roughly sizing and estimat-ing the price of a drying system. To obtain them, multiply the chart readings by the evaporation rate in lb/h.

The charts are based on typical evaporative drying conditions—pressure of 1 atm, supply air at 60°F and 0.007 lb/lb DA moisture, and feed at 60°F and 50 percent moisture. Heating is direct-fired with no. 2 fuel oil, and heat loss is 0.5 percent of dryer inlet enthalpy. Results are comparable to using the psychrometric charts without heat effects. But results are off significantly for large heat effects, such as lines *B* and *C* in Fig. 6.3.

With natural gas or indirect heat, the error in the airflow rate is less than 0.5 percent. To correct the heat load, add 4.0 percent for nat-ural gas heating and subtract 7.6 percent for indirect heating. Both

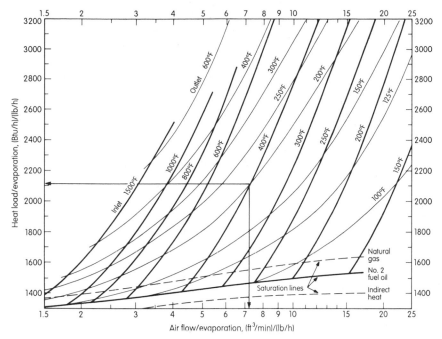

Figure 6.6 Dryer shortcut chart for temperatures to 1500°F.

values are roughly 9 percent higher at 80 percent feed solids and 9 percent lower at 20 percent.

The limits of dryer outlet saturation are identified by the three heat source lines—indirect heat, fuel oil, and natural gas. Values below the line for each heating method are beyond saturation. Natural gas combustion adds the most moisture to the airstream, so it restricts the usable area the most.

When interpolating values, note that the horizontal axis is logarithmic, and the air temperature lines are also spaced in a kind of logarithmic pattern. To obtain airflow and heat load, first find the intersection of inlet and outlet temperatures, without going below the saturation line for the heating method used. Read the value on each axis, and multiply each by the required evaporation rate in lb/h to get airflow and heat load.

As an example, a 400°F inlet and 200°F outlet has coordinates of 7.1 and 2110. Then an evaporation rate of 1000 lb/h requires an airflow of 7100 ft³/min and a heat load of 2,110,000 Btu/h. These agree with the values in Table 6.1, column III, except that heating by natural gas increases the heat load about 4 percent.

Figure 6.7 extends Fig. 6.6 into lower temperatures. It is based on indirect heating, however, which is more common for dryers operating

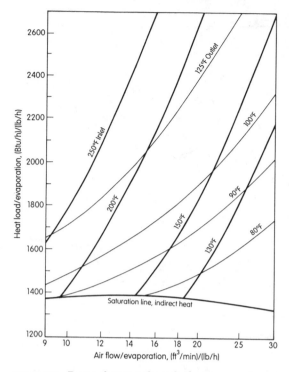

Figure 6.7   Dryer shortcut chart for low temperatures.

in the low range. Because temperature differences are much smaller in this area, heat effects such as feed temperature and heat loss have a much greater influence on the results.

The reader can add to these shortcut charts to further their usefulness, as in helping to estimate equipment sizes and prices.

7

Surveys

A survey is the first step toward reviving a drying system, and the main aspects are planning, taking data, calculating, analyzing, and making recommendations. The goal is to improve operations or to compare them with some standard, such as the original design. The methods described here have been refined over time as a result of having conducted over 100 surveys of drying systems, which led to subsequent corrective actions with regard to faulty conditions.

7.1 Purpose

A survey should ultimately result in better operations and higher profits, without sacrificing safety or commonsense practices. Systems should be made more energy-efficient and less prone to shutdowns. The main purposes served by dryer surveys are the following.

1. Identify the cause of operating problems.

2. Confirm equipment design or guarantee.

3. Study a new drying situation—a new product, revised operating conditions, or relocation of the dryer.

4. Optimize operating conditions to improve energy use and productivity.

5. Modify operating conditions for desired product quality.

6. Analyze the costs of proposed equipment changes and additions.

7. Help instruct operating personnel in correct procedures.

8. Check the accuracy of the system's instruments.

In addition, surveys often have unexpected benefits in uncovering situations that need correcting. These may include air leaks, product buildup, excessive pressure drop, temperature stratification, poor air distribution, collector inefficiencies, flow disturbances, and various other problems that penalize good operation.

Fundamentals. After the planning stage of a survey, measurements are taken—flow rates, temperatures, pressures, and moisture contents of the streams of air, feed, and product. Other data may be needed on the heat source, feed and product properties, and, with some systems, electrical measurements. In addition, dust samples may have to be taken at collectors to judge their efficiencies, or at the exhaust stack to comply with federal or local laws.

A complete survey also includes inspecting the equipment while shut down, checking especially for product buildup. Other functions are making calculations, analyzing the system, and recommending corrective action, usually in a written report.

Plant instrument readings should be verified, and it may benefit plant personnel to be instructed on how to take and interpret data. Instruction may also be given in making calculations and analyzing drying conditions to help assure continued operation at full potential.

7.2 Planning and Executing

When there are problems in a drying system, the company's personnel usually try to correct them, but may lack some of the expertise and the instruments needed for taking data. Also, those who can conduct a survey are usually experienced in several disciplines, but not in the detailed knowledge of drying. Thus they cannot analyze the data fully and draw the right conclusions. Many firms, therefore, engage the manufacturer or an engineering or consulting firm to handle either the analysis only or the total project. Plant personnel can assist, and thus be better able in the future to do some survey work and avoid problems.

Planning. The first step in a survey is to study the operating data and compare them with the original design. Because of improvements or technical or commercial changes, few plants continue to run the exact product for which the plant was designed. Even when the product remains the same, operating conditions may be different. Some changes affect drying results seriously, and the survey crew should be aware of them before they start.

The next step is to plan the survey with the plant manager and engineers, giving proper consideration to safety and environmental concerns. A review should be made of current and planned operations, past problems and their trends, the schedule of the survey, and what equipment changes may be necessary (such as drilling holes for probes). Access to the equipment must be provided as needed. The planning and execution are easier when a plant is not operating on a full schedule. If shutdowns can be arranged to suit the survey, there will be fewer delays.

Third, a working schedule should be prepared and details agreed to by all concerned. Inspection of the equipment requires shutdown, and, even if brief, it may alter work schedules for several company departments. All production shifts should be involved in this aspect and understand the purpose and benefits of the survey. It is also important that they be familiar with the work schedule, which should list all data items and when they are to be done, including the taking of samples. Any necessary forms should be ready before the work begins.

Figure 7.1 is a survey form for an energy-productivity study on a direct-drying system. It is intended mainly for data gathering by plant personnel, with the data analysis to be made by the user's engineers or an outside firm. An analysis of such data is presented in Chap. 8.

Conducting the survey. The key to conducting a survey is to get the information as quickly and completely as practical, and the trade-off between these two has to be watched. To get a set of data that is reliable and consistent, speed is essential. Otherwise changes in either plant operations or weather may nullify much of the work. Complete-

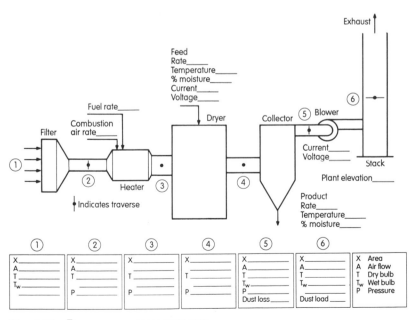

Figure 7.1 Dryer survey measurements.

ness is important, and some duplication is needed because certain readings may prove to be questionable and need confirmation.

Discussions with supervisors and operators can reveal possible snags. Operating logs may show the degree of consistency and accuracy in running the system and any past process upsets. Temperatures and pressure drops recorded by the plant instruments, and other data entries, are the first items to be checked to see whether the system is performing as it should.

Surveys are not conducted in any set order of readings and observations. The sequence varies with the type of dryer, application, plant size, and number of technicians. It also depends on the specific data items needed and the arrangement of equipment—the number of levels, in particular. For example, it is sometimes possible for a technician working alone to start at the top level and work down to minimize stair climbing. Or it may be better to start at the bottom and work upward to the reading at the stack, and then return to the lower level.

Near the start and end of each set of data, the heat source and product rate readings should be taken, as well as the feed rate when possible. Then the evaporation rate can be calculated and compared with results of other measurements. For each reading the time should be recorded. If results are not consistent, readings may have to be averaged or combined for an overall result. Knowing which data will not be available, or might be unsuitable, helps in planning the steps to a satisfactory set of data in the shortest time period. Allowances should be made for measurements that take a long time, such as traverse readings and vessel loading.

To conduct the measuring portion of a survey on a large dryer, it is preferable to have at least two engineers or technicians. One of them must be skilled in the use of the instruments. In addition, on any survey, regardless of dryer size, one of the technicians has to be familiar with the drying technology because decisions will have to be made as the work progresses.

A fairly complete survey on a typical installation will take a minimum of 1 day; 3 to 5 days are more likely for a large installation. Depending on the condition of the system and on the cleanliness inside the ducts and the equipment, the system may have to be shut down for 2 or 3 h. Taking an additional day is typical for calculations, analysis, and report writing.

Summary of measurements.   Table 7.1 lists the data normally taken in a full survey on a direct dryer. Items omitted include those required for a specific dryer type, and some that may occasionally be needed

TABLE 7.1   Summary of Survey Measurements

Heat source
  Type (steam, natural gas, fuel oil, electric, other)
  Heat rating (pressure, if steam)
  Flow rate
  Exhaust temperature (for indirect heater)

Feed
  Type (solution, slurry, paste, wet granules, sludge, press cake, etc.)
  Temperature
  Flow rate
  Percent solids (or percent moisture)
  Particle size
  Angle of repose
  Other properties (e.g., viscosity, pH, specific gravity, bulk density)
  Voltage and current for feeder drive and agitator or shell drive

Product
  Temperature
  Flow rate
  Percent moisture
  Buildup on equipment walls (locations, amount, and dryness)
  Particle size
  Bulk density
  Other agreed-upon product characteristics (e.g., color, taste, aroma, flow properties)

Air
  Dry-bulb temperature at heater inlet, dryer inlet, dryer outlet, stack inlet
  Wet-bulb temperature at heater inlet, dryer outlet
  Flow rate at heater inlet, dryer outlet, collector outlet(s), stack inlet
  Pressure difference across dryer, collector(s), blower(s)
  Voltage and current of blower(s)
  Dust loading at collector outlets and stack inlet

when troubleshooting. Tests for feed and product are often conducted by plant personnel, as in a control laboratory, and this provides the highest accuracy. Readings from fast-result instruments, such as moisture balances located at the dryer, are in many cases only approximate.

It is assumed that any needed equipment dimensions can be found on drawings; otherwise measurements must be taken and recorded.

For some dryers additional data may be needed for calculating exposure time, percent solids loading, air moistures, and other values, and to locate and calculate leaks. Also, depending on the type of dryer, samples may have to be taken to determine feed specifications and product characteristics.

Instruments.   Instruments mentioned in this text have proven suitable for the accuracy required, and they are relatively easy to use. They are also commonly available (except for an air-sampling device that can be constructed by any good machine shop). Various other instruments have been found wanting—less reliable, too expensive, less universally applicable, in need of frequent calibration, not sufficiently portable, or more time-consuming to use. This does not infer, however, that there are no others suitable, at least under some circumstances.

On-line instruments for continuously recording temperatures and flow rates are very convenient. Many need periodic calibration, however, which tends to be neglected. Thus the plant's instrument readings should be verified first—if not correct, they may affect some of the planning decisions and the rest of the survey schedule. These checks are made with portable instruments, which are less convenient but have the needed flexibility.

Portable thermometers for use in dryer surveys are generally of the thermocouple type. Glass thermometers are unsafe, thus banned by many firms; mercury is taboo at food and some other plants. Dial or digital display types are in common use, and accuracy to the nearest degree Fahrenheit is best. Greater sensitivity—often with wild fluctuations that are more confusing than helpful—would not be consistent with the flow rate and other data, particularly when using psychrometric charts.

Probes and traverses.   When a survey is run on a system for the first time, small holes must be drilled for airflow, pressure, and temperature probes. After the job has been done, the holes should be plugged with cork or neoprene stoppers, or covered with a tape material that withstands the operating conditions.

Readings in ducts should be taken in the longest straight length available, where turbulent or spiral flows have smoothed out to some extent. Unfortunately the recommended straight length of at least eight diameters—preferably more—on both sides of the probe is rarely found in drying systems. Thus the highest accuracy cannot be expected. In any case, the longest possible straight run of duct before the measuring point should be used. An alternative for greater accuracy is to install straightening vanes.

Multiple traverse readings for airflow, and sometimes for temperatures, are needed in all but the smallest cross sections. This permits averaging out the effects of turbulence on airflow readings and the effects of stratification on temperatures.

One traverse method for a circular cross section is to consider it divided into concentric rings of equal areas. For duct diameters up to 2 ft (61 cm), four rings has been found to be adequate. On each ring two sets of four readings are taken at right angles to each other. Rectangular cross sections can be divided into nine equal areas, arranged three by three, but other patterns may be needed for ducts not close to square, or very large, and for very turbulent flows.

7.3 Feed and Product Measurements

Of the various methods for determining feed and product properties, some are unique to one specific product, a few are proprietary, others are designed to be used on a class of materials, sometimes specified by industry associations. Because of this diversity, tests on product and feed are, in general, not performed by production departments. But many plants have one or more instruments at the dryer. Most common are moisture balances, particle size screens, and density meters to provide quick results on properties that are critical to the drying process.

Changes in the feed are almost sure to affect drying results. Feed prepared in batch operations may differ from one batch to the next. Even if the preparatory operation is continuous, properties that affect the drying can vary. For this reason, the relevant feed properties should be checked on a regular basis.

Product temperatures.   The temperature of particulates is usually difficult to measure accurately because of the inclusion of air and the cooling effect of any moisture. An accepted method is to place a sample of the dried product in an insulated container and wait for equilibrium. Product from short-exposure-time dryers may take several minutes to equilibrate in both moisture content and temperature.

Feed flow rates.   A variety of accurate meters are available to measure liquid flow rates, but because of their expense and the low priority given to measuring feed rates, they are seldom installed. Rotameters and vane-type meters are used occasionally, but are limited to solutions, and some need frequent calibrating.

The flow rates of liquids or wet solids can be determined by weighing methods. For either solids or liquids scale-mounted tanks with load cells give accurate results. On the other hand, level difference measurements are inaccurate, especially in large-diameter tanks (which are not always accessible). If the tank has an agitator, shutting it off may be impractical.

Moisture contents.   At many plants feed and product moisture contents are the most important properties from the standpoint of dryer operation. Feed moisture has a major effect on dryer capacity, and it may be quite different from the value on which the system was designed. If it tends to vary, it should be checked often. The effect of feed moisture on productivity is detailed in Chap. 8.

Particle size and size distribution. Properties that may have to be measured—because sometimes they have a major influence on dryer operation—are product particle size and size distribution. Methods vary widely in sophistication—visual inspection (usually by microscope with a sizing grid), sifting through sets of screens, and using instruments that employ a light-scattering technique and computer analysis. Sets of screens in graduated sized openings are in common use, and are vibrated or air purged to speed results. All these methods, visual inspection in particular, require a carefully prepared representative sample.

Bulk density. Bulk density affects dryer loading in bed-type dryers, and affects packaging for nearly all dryers. Particle density and absolute, or material, density are seldom of concern, but are important for certain products. The bulk density is measured by a number of methods to suit specific materials. The most common is simply to fill a graduate that has been cut off at 100 ml, and then to determine, by weighing, the *as poured*, or *loose* density. Then a *settled*, *tapped*, or *packed* density is found by tapping the graduate either a fixed number of times or until settling stops.

Other properties. A significant property, but one often overlooked, is the heat of crystallization, discussed in Sec. 5.2.3 under *Other heat effects*. Because its value may be positive or negative, it can add or subtract heat from the drying process. If its value cannot be found in the chemical literature, it can be measured using a calorimeter. A rough check of the value and its sign can be made by dissolving a measured sample of the dry material in water and comparing any temperature change against that of a material with known heat of solution. The heat of reaction has the same effect on drying conditions, but is not often encountered.

Other feed properties that can affect dryer operation are composition, stickiness, toxicity, particle size, consistency or viscosity, pH, and temperature into the dryer. Another, vital to success in some bed-type dryers, is the feed's angle of repose or other gauge of its ability to flow. The survey technician is not likely to have to measure such properties.

7.4   Heat Source Measurements and Heat Losses

In many plants the flow rates of heat sources—most often steam, natural gas, or fuel oil—are measured by flow meters. But these meters often measure the flow to other equipment in addition to the dryer.

The other uses have to be shut off to get the proper reading for the dryer. Flow rates of dryer heat sources sometimes vary widely from fluctuations in control, and thus averaging a number of readings may be necessary.

With steam heating the dryer heat load can be figured from the condensate by either metering or weighing. If the condensate is released to the atmosphere, the flash has to be calculated and added to the condensate—at sea level it is about 24 and 34 percent for steam at 100 and 150 lb/in^2 (689 and 1034 kPa), respectively. At higher elevations those percentages are higher. Because condensate flow is often irregular, several readings must be taken and averaged if not consistent.

When heating indirectly, including by steam, the exchanger that heats the airstream has a radiant-convective heat loss. The loss is a function of the heater's surface area, its inside and outside surface temperatures, and its insulation rating. It can be estimated using insulation manufacturers' data as given in App. B and will normally be much less than the traditional recommendation of 5 percent.

An additional and much larger loss results when a fired indirect heater is used. This loss can be reduced considerably, however, by partial heat recovery of the heater exhaust. Because of the heat losses, the heat input to the dryer is best figured from the airflow and temperatures of supply air and inlet air to the dryer. The heat input can be calculated using the enthalpy difference, as discussed in Sec. 5.2.8 under *Calculating leaks and auxiliary airflows*.

A heat loss from an insulated drying vessel of 5 percent of the heat load is a traditional value for calculating airflow, but one that can hardly be justified. It is roughly 4 to 10 times higher than the true value for a typical dryer, and it can affect the airflow rate by 15 percent or more. Based on heat load, the heat loss even from a vessel with a large surface area is typically only 0.2 to 0.8 percent. See also Sec. 5.2.3 under *Heat loss and safety factors*.

7.5 Air Measurements

Measuring air conditions is more difficult in large commercial equipment, and data are valid only after a steady state has been reached. With oil heating, equilibrium takes longer, and much longer with refractory-lined heaters. Readings should not be taken until a steady state has been reached. All readings and significant observations should be documented. Without a complete record, an otherwise good survey is worth much less, especially when things that seemed obvious at the time can no longer be remembered.

A summary of data items is often given on a diagram such as Fig. 7.2, which illustrates typical airstream measurements and locations

COMPANY NAME _____ TELEPHONE NO. _____ DATE _____
PLANT _____ PLANT ELEVATION _____
ADDRESS _____ OPERATING HRS/YR _____
PERSONNEL _____

OPERATING DATA

DRYER TYPE AND MANUFACTURER _____.

HEATER: TYPE _____ INLET AIR TEMPS: DRY BULB _____ WET BULB _____.

HEAT SOURCE: TYPE _____ RATING _____ FLOW RATE _____ COST _____

ELECTRICITY: FAN MOTOR HP_____ AMPS _____ VOLTS _____ COST _____
             FEED DRIVE UNIT HP _____ AMPS _____ VOLTS _____

FEED & PRODUCT: NAME _____
       CONSISTENCY _____ HEAT OF CRYSTALLIZATION _____
       FLOW RATE (FEED   PRODUCT   EVAPORATION) _____
       FEED: SOLIDS _____% ._____MAX.% TEMP. _____MAX _____
       PRODUCT: MOISTURE ____% _____MAX.% TEMP. _____

DRYING VESSEL: DIMENSIONS _____
       AIR TEMPS.: INLET _____ OUTLET: DRY BULB _____WET BULB _____
       INLEAKAGE _____
       INSULATION: THICKNESS & COVERAGE _____

PRODUCT COLLECTORS: TYPES, MANUFACTURERS, PRESSURE DROPS

_____
_____

AIR FLOW	FLOW RATE OR VELOCITY & DUCT AREA	TEMPERATURES DRY BULB WET BULB	
HEATER INLET	_____	_____	_____
DRYER OUTLET	_____	_____	_____
COLLECTOR IN	_____	_____	_____
STACK INLET	_____	_____	_____
OTHER _____	_____	_____	_____

COMMENTS _____
_____
_____
_____
_____

NOTE: Provide as much data as possible on each item, so that a full
energy and productivity analysis can be made. Indicate the specific
U.S. or metric units. Add any relevant comments, e.g. on data ranges or
accuracy, influencing factors, etc. Temperature recorder readings are
often inaccurate; confirm with thermometer or thermocouple. Make
multiple temperature and air flow traverses across large ducts,
especially needed for high temperatures. If possible, locate probes for
instruments after a long section of duct.

Figure 7.2 Survey data sheet for energy and productivity analysis.

for a drying system. Note that some data are not taken in hot air, while others are taken only in clean air before the dryer and after the collector.

Air temperatures. In measuring the temperature of a fast moving airstream, readings within about 5 percent of each other are generally considered to be satisfactory. A 10 percent difference is at the bounds of acceptance. But both the needed precision and the accuracy depend on the circumstances. The importance of accuracy increases, for example, as the difference between the dryer inlet and outlet becomes smaller. This was illustrated in the survey of a large, very low temperature flash dryer. A change of 2.0°F (1.1°C) in either inlet or outlet temperature affected both energy use and production rate by more than 10 percent.

If an air temperature recorder cannot be calibrated electrically, it has to be checked with a portable instrument. To do this, the dryer is operated manually, and the recorder's sensor removed from its well and replaced with a calibrated thermometer. There should be good agreement between these readings and with the average of traverse readings taken nearby. If not, the well should be reset in a location that gives greater accuracy. In airstreams near the heater, the sensor may have to be shielded against radiation effects.

The temperature difference and airflow rate measured between two points can be used to calculate radiant-convective heat loss, if conditions are otherwise isothermal. But if there is a blower in that section, allowance has to be made for heat added at about 0.5°F/in WG (1.1°C/kPa).

Ambient wet-bulb temperature readings can be taken with a sling psychrometer. But for a stream of warm air in a duct, a more elaborate method is needed. The wick should be wet with water (preferably distilled), warmed to the wet-bulb temperature or slightly above it. The reading is taken when the value remains constant or dips to a minimum. The general uncertainty of wet-bulb determination often requires extra readings to get acceptable replication.

In a dryer's hot air inlets and in ducts conveying powder, velocities are usually in the range of 2000 to 5000 ft/min (10 to 25 m/s). Except at the lower end, these velocities are high enough to spoil the accuracy of wet-bulb readings. In conveying ducts wet-bulb readings are often invalidated when powder forms a crust on the wick. Sometimes readings can be taken quickly enough to avoid the problem. (A prewarmed thermometer saves time.)

To overcome either problem, a device may have to be used that takes out a sample stream of the air at an ideal velocity—about 2000 ft/min (10 m/s)—and removes the powder with a filter or tiny cyclone. For this to give the right wet-bulb reading, the dry-bulb temperature must be held constant. As a last resort to solve the powder problem, the dryer can be operated evaporating only water (instead of feed). But dryers cannot always operate on water at the same conditions as with the feed.

Airflow rates. The airflow rate should be determined at selected points to verify the system's volumetric airflow. These multiple readings are also a means of calculating the evaporation rate, described in Sec. 7.6. The airflow rate can be found by several methods, and because replication is difficult, using more than one is advised. These include the following.

1. Measure velocity pressures with a Pitot tube and pressure gage or manometer, and compute velocities, then airflows using the cross-sectional area.
2. Measure the cyclone pressure drop, converting to airflow using the manufacturer's data.
3. Measure the fuel rate and convert it to heat load and airflow using the heat rating of the fuel and enthalpies of the air. This method is useful for high air temperatures.

Pitot tubes are used with pressure gages or manometers to read velocity pressures. For pressure differentials below about 2 in WG (0.5 kPa), low-range diaphragm pressure gages with a direct reading of velocity are the most convenient. U-tube manometers are at least as accurate—inclined types are even more so, and a low-density fluid gives still greater sensitivity. In many field applications manometers take time to set up and level.

Velocity pressures can be converted to velocity and airflow rate using the instrument manufacturer's instructions or handbook equations (Jorgensen, 1983). If the readings are very widely scattered, a root mean square calculation can be made, but an arithmetic average is nearly always sufficiently accurate. In a round duct, taking a reading near the center and multiplying it by 0.9 gives a quick, but very rough approximation.

Pressure drop readings across cyclones have in most cases given satisfactory airflow results. Manufacturers' charts generally relate pressure drop to air velocity or directly to volumetric airflow. When a cyclone becomes partially plugged, however, the flow rate is reduced, and the readings fluctuate, sometimes wildly.

Determination of the airflow rate by weight from the heat source is given in Sec. 5.2.8 under *Calculating leaks and auxiliary airflows*. If the airflow rate is measured before entering a direct-fired heater, the combustion air must be added to get the total airflow into the dryer. Leaks between the measuring point and the dryer inlet also have to be considered. Because leaks are so commonplace, this method is less reliable and best performed as a last resort, or as a check on more direct measurements that for some reason are not trusted.

The methods described are generally satisfactory, but there are others. Venturi meters and other pressure-difference devices would be useful on the clean-air sides of drying systems, for instance, but are seldom installed. Fan curves can be used as a rough approximation, but the results are not reliable. Anemometers are limited to cool, clean air, and some need calibration before each use.

Leaks.   Most dryers have draft-type airflow so that leaks, if any, are inward. Leaks are hard to detect and even harder to determine, usually from flow rate differences. However, large leaks, which may be deliberate for cooling purposes, can be estimated by accurately measuring airflows at the right places. Pressure differentials across collectors are often the easiest method.

Merely locating small leaks is difficult if the opening is not obvious, such as under insulation. Sometimes leaks can be found from the noise they make. They can also be detected visually using smoke or fine powder near suspected locations. Fumes from dry ice or fumed silica can also be used. Section 5.2.8 describes the calculation of leaks from airflow measurements.

Air pressure drops.   Pressure drop should be measured across the drying vessel, across all product collectors, and at any other critical locations. These readings can be taken at the same time as the airflow measurements using either a diaphragm gage or a manometer.

Powder loadings.   Powder or solids loading is often important, even though not a major factor in dryer capacity. It may be needed to assess product loss or collector efficiency, or to evaluate compliance with federal or local pollution laws. Powder loading is the weight of solids per volume of air, usually measured in gr/ft^3 or mg/m^3.

The powder sampler shown in Fig. 7.3 has been developed to measure loading and losses, and has been used for several years with good results. Inserted into a conveying duct or stack, it filters out the solids, enabling the loading and the losses to be calculated. It requires a 2-in

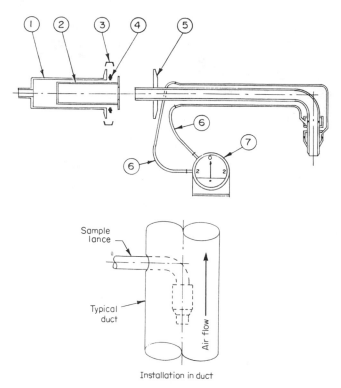

Figure 7.3   Powder sampler. 1—canister; 2—felt or paper dust bag; 3—clamp ring; 4—gasket; 5—sample tube and flange; 6—hoses; 7—diaphragm pressure gage.

(5-cm)-diameter opening in the duct, preferably where the airstream is not turbulent.

A stream of air is passed through the sampler tube, and the same ratio of solids to air is maintained in both tube and duct by keeping the same velocity in both. The pressure drop of the tube and filter is overcome by pulling air through the sampler, and controlling it, by use of a plant vacuum system or portable vacuum cleaner. Equal velocities are assured by keeping equal static pressures in both tube and duct, thus a reading of zero on the diaphragm pressure gage.

The volume of air passing through the sampler can be computed using the sampling time, the tube cross-sectional area, and the velocity in the tube. The powder loading is the weight of solids over the volume of air. The rate of solids loss is the sample weight times the ratio of duct-to-tube areas divided by the test time.

Readings can be taken at a single point known to give the correct average, or a traverse can be run similar to those for temperature and airflow. The time required to accumulate a representative sample of powder varies with the dust loading in the duct or stack, but is typically 20 or 60 min when taken downstream of a cyclone or bag filter.

7.6 Calculation of Evaporation Rate

The product rate should not be used as the only gauge of dryer performance. Except for high solids content feeds, as in bed-type dryers, it has little effect on capacity. The evaporation rate, conversely, is a major influence. Thus the production rate for each unit of evaporation P_r/E_v is a vital ratio. In the following equations E_v, F_d, and P_r are evaporation, feed, and product rates, and F_s, F_m, and P_m are feed solids, feed moisture, and product moisture on a percent wet basis.

$$E_v = F_d - P_r \tag{7.1}$$

$$\frac{P_r}{E_v} = \frac{F_s}{100 - F_s - P_m} \tag{7.2}$$

$$\frac{P_r}{E_v} = \frac{100 - F_m}{F_m - P_m} \tag{7.3}$$

Equation (7.2) shows that evaporation and product rates are directly proportional to each other. The effect of feed solids on the two rates is nonlinear. As feed solids increase, the ratio is increased by the rising numerator and falling denominator. Although product moisture has less influence, it becomes important rapidly with rising feed solids at the high end. Thus if productivity is to be kept at its best level, the evaporation rate should be determined at frequent intervals, along with the feed solids.

The evaporation rate has to be calculated because it cannot be measured. The simplest way is to measure feed and product rates and use Eq. (7.1). If the rates fluctuate from causes outside the drying system, data should be taken at peak conditions. The product can be weighed as discharged and averaged over a reasonable period, taking into account shutdowns.

The product rate includes losses, buildup, and cleaned-out and rejected material (these may have to be estimated), but it does not include product recycled to the dryer, because the latter passes through without further evaporation. Feed rates are more difficult to determine, and most plants do not measure even liquid feed rates on-line. Thus the product rate generally has to be used with Eq. (7.2) or (7.3).

When conducting a survey of a newly installed dryer, often the feed is not available at full rate. In such a case the dryer can be run at reduced feed rate and with reduced inlet air temperature to satisfy the rate, but at full airflow rate and with everything else at design conditions. The actual airflow and evaporation rates are determined as described in Sec. 7.5. Then the increased evaporation rate is calculated for the design inlet air temperature, using either the Chap. 5 or the Chap. 6 method. It can then be compared to the design rate.

For direct-fired units another path to the evaporation rate is to calculate the outlet airflow rate A_3, as mentioned in the section on airflow rate. Using A_3 from one or more of these methods, the evaporation rate is

$$E_v = A_3(M_3 - M_2) \tag{7.4}$$

where M_3 and M_2 are the air moistures at dryer outlet and inlet. Because this method is indirect, the results may be less accurate.

8

Improving Operations

8.1 Introduction

It is often said that energy costs less to conserve than to purchase. Yet many drying plants continue to buy excess fuel, running for years at the conditions that had been approximated during the haste of the original test work. At one major testing laboratory the final inlet air temperatures were set at even multiples of 100°F for more than half of all the tests. Most of the other tests were at low temperatures, with the inlets set at an even 50°F.

Another limitation of dryers is that they are not designed and built to exacting standards. Thus the possible improvements in efficiency range from simple corrections to complete system redesign. At some plants the easy changes have been made. So to enhance operations further, a procedure is needed to find and evaluate the most cost-effective options in both existing dryers and those being planned.

There are four basic strategies to improved dryer performance, and each has several options. They apply mainly to dryers using air for heating; purely indirect dryers have less potential for improvement.

1. Reduce the evaporation or make it more efficient.

2. Optimize heating temperatures and air moisture contents.

3. Reduce or partially recover heat losses from the dryer surface or the exhaust.

4. Provide heat to the dryer by some means other than by heating the supply air.

Most steps taken to save energy also increase the rate of production, which may be of even greater benefit. An existing dryer can produce more; one that is being planned can be made smaller. On the negative side, most energy-saving steps increase the dryer outlet humidity, and, depending on the product, this can limit the extent to which changes can be made. It may also require other measures, such as the use of a secondary dryer. For a discussion of outlet air humidity see Sec. 5.2.8.

In general, those steps that reduce heat also reduce electric power use. In nearly all cases power costs much less than heat, but power savings can be more than 20 percent of the total savings, even more when the airflow is substantially reduced, as with two-stage drying. However, power is increased by some heat-saving steps. A prime example is exhaust-to-supply air heat exchangers, which have great potential for heat saving, but increase system pressure drop.

When dryer tests are run, the projected costs of the proposed drying system and its operation become increasingly important. There is a better chance of success (or stopping losses quickly) if estimated costs for the system under study are compared continually to a product cost standard while the tests are in progress. The costs can be computed using spreadsheet techniques and can be made a part of the test program.

8.2 Indirect Dryers

Because they use little or no air, indirect dryers are more energy-efficient than direct dryers and benefit less from energy-saving strategies. Their diffusion dominated applications can be improved little or not at all. They may, however, serve as secondary dryers in multistage systems.

Heat transfer dominated applications can benefit by increased transfer of heat. In bed-type dryers improvement can come by either reduced fouling or better mixing, both of which are a function of the feed properties and agitator action. When permitted by the application, agitation should be vigorous in local areas with no short-circuiting of wet particles to the discharge. This kind of efficient mixing depends on the agitator design and the speed of rotation. The design should allow the maximum speed, without uncovering or fouling the heat transfer surface or causing fluidization, all of which reduce the heat transferred. Effective heat transfer also depends on residence time-temperature relations that give each particle equal treatment. The direct dryer types vary in the degree that they achieve these ideals.

The other part of the heat transfer equation, Eq. (5.6), that can be influenced is the temperature difference. Improvement is restricted, however, by the need to heat the solids above the liquid's boiling point, but not above the solid's degradation temperature. Zoned heating sections are sometimes cost-effective because they allow some control, such as higher temperature against the wet feed. Adding heat to the bed by backmixing in hot product may help the heat transfer rate, but it adds to the required bed volume.

8.3 Direct Dryers

The greatest potential for energy savings and productivity improvement in direct dryers is reducing their demand for airflow. It is necessary, however, to keep above any minimum needed to fluidize or convey the solids. These requirements and the evaporation rate and heat load are related to the weight flow of air. But it is the volume flow that affects the capacity, size, and cost of the dryer and its components. Thus both weight and volume flow influence the improvements.

Table 8.1 summarizes the percent fuel and power savings and production increases for some of the simpler changes. The amount of each change has been chosen as practical for a direct dryer with moderately low operating conditions—25 percent feed moisture and inlet and outlet air temperatures of 300°F (149°C) and 170°F (77°C), respectively.

Each of the results in Table 8.1 is independent of the others, and multiple changes are not additive. The correct total for more than one change requires that a calculation be made that includes all changes.

Some optimizing involves system components other than the drying vessel and potential problems in them. For instance, heat recovery is increased in an exhaust-to-supply air heat exchanger, but at a higher pressure drop and at the risk of fouling and corrosion.

8.4 Implementing

Energy and productivity improvements are implemented by first obtaining accurate operating data from either a survey or test data. Then calculations are made to show the effect of operating or design changes on the drying results. Surveys and calculations are detailed

TABLE 8.1 Effect of Simple Changes on Direct Dryer Performance

Condition	Change	Fuel saving, %	Power saving, %	Production increase, %
Inlet air temperature	+10°F	14	16	16
Outlet air temperature	– 5°F	14	14	14
Feed temperature	+50°F	10	10	10
Feed moisture	– 1%	16	16	16
Leakage	– 5%	4	10	9
Insulation	2 in	5	5	4

in previous chapters. The third step, described here, is to determine the cost benefit or penalty for each of the changes, using assumed operating hours per year and costs of heat and power. Analyses can be carried further if other production costs are known together with the product's selling price. Computer spreadsheets are a convenient analysis tool in this work.

This survey-calculation-analysis technique has been applied to a number of existing direct dryers to improve manufacturing margins. It can also aid in designing new equipment, as well as in optimizing test conditions or evaluating the relative operating costs of different dryers. Operating changes on plant dryers are best made using evolutionary operations techniques. This entails making gradual changes while monitoring the product specifications. Then the maximum benefit can be gained without upsetting dryer conditions or production schedules.

Before starting an analysis it is important to know the effect of changes on operating conditions and what else they may influence. To illustrate, product moisture, unlike feed moisture, is not usually considered to have an important effect on drying results. Often it is only kept within a relatively broad range to meet market requirements. But if the feed moisture is low, the product moisture has an important bearing on productivity. The use of Eq. (7.3) will show, for example, that at 10 percent feed moisture a dryer will have a production rate that is 12.5 percent higher at a product moisture of 2.0 percent as compared to 1.0 percent.

In a dryer heated by air, aiming for the highest permissible product moisture also allows higher outlet air moisture, which gives better drying efficiency. If lower product moisture is desired, but a higher outlet air temperature would harm the product quality, the inlet air temperature can be lowered. This lowers the production rate, but it also lowers the outlet air moisture, and thus it can lower the product moisture. It may be necessary, however, even though it hurts drying efficiency.

Spreadsheet calculations were generated to illustrate the technique of analyzing energy and productivity improvements for direct dryers. Using the conditions in Table 8.2, three sets of calculations were made, each using different inlet and outlet air temperatures—(a) low, (b) medium, and (c) high. The evaporation rate of 10,000 lb/h (4536 kg/h) was selected to result in a dryer of commercial size for all temperatures.

Seven energy-productivity options were each altered in turn and the effect calculated. Heat loss and product temperature were computed for each arrangement. The equations needed for the spreadsheet cells

TABLE 8.2   Data for Sample Calculations in Table 8.3

Evaporation rate	10,000 lb/h
Feed solids	30% by weight
Fuel	Natural gas
Feed temperature	60°F
Supply-air temperature	60°F
Supply-air moisture	0.007 lb/lb DA
Heat of crystallization	0.0 Btu/lb
Fuel cost	$6.00/MBtu
Power cost	$0.06/kWh
Operating rate	6000 h/yr
Product moisture	5.0 wt %
Solids specific heat	0.4 Btu/(lb · °F)
Collection efficiency	100%
Safety factor	0.0%
Elevation	Sea level
Pressure drop	16 in WG
Pressure	1.0 atm
Leakage	None
Insulation	2.0-in calcium silicate

are relatively simple and logical, except those for electric power, which have to be derived for the major drives of the specific dryer type. The spray dryer design was used, for which the power is higher than for most others.

Results. Table 8.3 lists the results of the three sets of conditions. All improvements and savings are shown as positive values, while effects made worse are negative. Set (*a*) is for an inlet air temperature of 300°F (149°C). The effects of higher air temperatures are shown in sets (*b*) and (*c*). These are for air at 500 and 900°F (260 and 482°C), respectively. Only the conditions, changes, and percent total savings are given for those two sets.

Figures 8.1 and 8.2 summarize the energy savings and productivity gains for 11 options. The first seven are those of Table 8.3. All are changed to the same extent and use the same numbers. Operating conditions were at the intermediate air temperatures, and feed solids ranged from 10 to 90 percent, so that the options would apply to all dryer types.

The options range from simple operational changes to complete system redesigns. As the energy savings in Tables 8.2 and 8.3 and in Figs. 8.1 and 8.2 show, some large gains come from changes that cost little. Air inlet and outlet temperatures, for example, can be optimized by a mere change in operation. In other cases it would be advisable to determine first by analysis the point at which added costs would be

TABLE 8.3 Analysis of Dryer Options

	Original design	1 Raise inlet air temperature, °F	2 Lower outlet air temperature, °F	3 Raise solids in feed, %	4 Raise feed temperature, °F	5 Reduce air leakage, %	6 Preheat inlet air, °F	7 Two-stage drying T; $M*$
			(a) Low-Temperature Inlet Air					
Design condition	—	300	170	30	60	5	60	170; 5
New condition	—	320	160	32	160	0	134	145; 10
Amount of change	—	+20	-10	+2	+100	-5	+74	-25; +5
Resulting data								
Heat required, MBtu/h	23.275	21.813	21.493	21.171	20.771	22.244	16.178	17.754
Airflow, actual ft³/min	98,943	86,112	90,236	89,990	88,757	89,854	98,943	73,132
Outlet relative humidity, %	15.60	17.30	20.70	15.60	16.80	16.60	14.80	32.20
Specific volume, ft³/lb	16.27	16.31	16.03	16.27	16.30	16.29	16.25	15.67
Pressure drop, in H_2O	16.00	12.08	13.50	13.23	12.85	13.18	20.03	9.07
Feed and fan power, kW	367.93	295.67	321.86	312.30	309.77	315.81	442.23	240.11
Production increase at design airflow, %	—	12.97	8.80	9.05	10.29	9.19	0.00	26.09
Energy reduction, %								
Heat	—	6.28	7.66	9.04	10.76	4.43	30.49	23.72
Power	—	19.64	12.52	15.12	15.81	14.17	-20.19	34.74

Energy savings at design production, $/yr

Heat		52,632	64,152	75,744	90,144	37,116	255,492	198,756
Power		26,015	16,586	20,026	20,940	18,764	−26,747	46,016
Total		78,647	80,738	95,770	111,084	55,880	228,745	244,772
Energy cost								
Total cost, ¢/lb product	3.504	3.220	3.212	3.158	3.103	3.302	2.678	2.620
Total savings, ¢/lb product	—	0.284	0.292	0.346	0.401	0.202	0.826	0.884
Total savings, %		8.105	8.320	9.870	11.448	5.759	23.573	25.225
Total heat, $/yr	837,900	—	—	—	—	—	—	—
Total power, $/yr	132,456	—	—	—	—	—	—	—

(b) Medium-Temperature Air

Design condition	500	230	30	60	5	60	230; 5
New condition	520	220	32	160	0	186	200; 10
Amount of change	+20	−10	+2	+100	−5	+126	−30; +5
Total savings, %	3.380	4.202	9.512	10.697	3.900	23.309	18.853

(c) High-Temperature Air

Design condition	900	280	30	60	5	60	280; 5
New condition	920	270	32	160	0	250	245; 10
Amount of change	+20	−10	+2	+100	−5	+190	−35; +5
Total savings, %	1.035	1.994	9.251	10.116	2.039	18.439	14.105

*T, M—outlet air temperature (°F); product moisture from first stage (%).

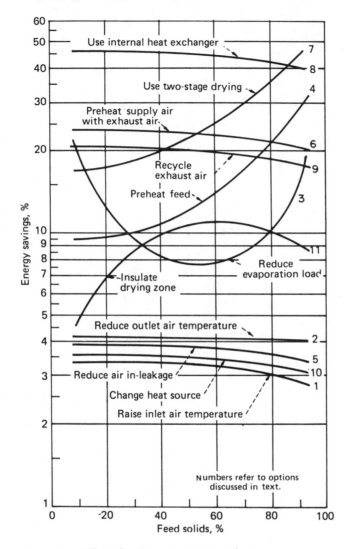

Figure 8.1 Effect of options on energy savings.

justified. For instance, before expenses are incurred for a feed heater, find the savings for heating the feed. Evolutionary operations or other techniques may be needed to find the best point for some of the options.

8.5 Improvement Options

The 11 energy-saving options are discussed in this section. Most of them also provide productivity gains. The last number of each subsec-

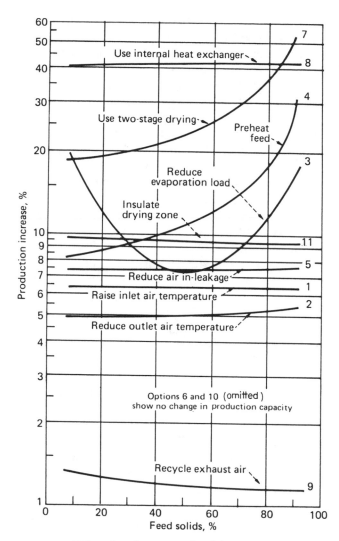

Figure 8.2   Effect of options on productivity.

tion corresponds to the numbers used in the tables and figures. Only the first seven appear in the tables.

8.5.1   Raise the inlet air temperature

A study of the psychrometric chart of Fig. 6.1 will show that the highest inlet and lowest outlet air temperatures allow the air to pick up the most moisture. Finding and maintaining this optimum ΔT is cost-effective, especially for dryers having low ΔT. Because evaluation tests are generally run with inlet temperatures stepped at large increments, it may be possible to raise the inlet temperature 10 to 70°F (6 to 39°C) without affecting the product. The outlet air humidity is only slightly affected. Each degree of increase in the air inlet temperature is less beneficial than a 1° decrease in outlet temperature. The inlet can usually be changed more than the outlet, but this advantage diminishes at higher levels.

For reasonably efficient operation of a commercial dryer, the air inlet temperature should be within 10 to 20°F (6 to 11°C) of its maximum. It is limited by the product's heat sensitivity or by materials of construction for the burner and hot air inlet section. When the inlet temperature is set higher, to maintain the same product properties, the outlet temperature may also have to be set higher. This higher outlet temperature to some extent affects benefits. See the discussion of the effect on air humidity in Sec. 5.2.8.

Particles suspended in a cocurrent airstream are kept relatively cool by evaporation. So for this mode, which is the most common, the inlet air temperature can be much higher than the product melting or degradation point. However, if powder clings to equipment surfaces, thus extending its time in the hot air, it may scorch and mandate a lower air temperature. Other considerations, such as product density and color stability, may set their own limits on either the inlet or the outlet temperature. In addition, the tendency of fine particles to scorch may force setting the inlet temperature at a point of low efficiency. It helps if the fines can be removed, as from flash dryers, and not overexposed.

A 20°F (11°C) increase is assumed for the examples of the tabulations and charts. The benefits are modest for this rise. Even so, close supervision and more accurate control of both inlet and outlet temperatures will often be cost-effective. In some cases computer control may be justified to maintain the optima.

8.5.2   Lower the outlet air temperature

The lowest practical setting of the outlet air temperature provides the greatest benefit to energy use and productivity, and usually to product quality. A unit volume of the cooler outlet stream contains more air by weight than does the warmer inlet. Thus its temperature has a greater influence on the energy use and productivity per degree of change. The benefit is actually provided by the higher moisture content of the air. It also imposes a penalty—higher air moisture may restrict the temperature reduction if the product moisture increases too much.

For the examples, a 10°F (5.6°C) decrease was assumed. The benefit for both energy cost and productivity is over 8 percent at the low-temperature condition, but less than 3 percent at the high temperature.

With some substances, and in certain dryers, a longer exposure time in the drying zone permits a lower outlet temperature, without increasing the product moisture. Release of hard-to-remove moisture sometimes needs more time. Occasionally a more uniform particle size can eliminate the need for either longer time or higher outlet temperatures, both of which may hurt the product quality or drying efficiency. If the outlet is set for proper drying of the largest particles, two-stage drying may be the most effective system design.

In some applications, unusual strategies are effective. Examples are two-stage drying for heat-sensitive products and control of the outlet air temperature by sophisticated product moisture instrumentation. The gain from allowing the product moisture to increase a little is small, except for very low moisture feeds. But it permits bringing down the outlet air temperature, and that may be very beneficial.

8.5.3 Reduce the evaporation load

Increasing the feed solids content seems so obvious as to need no comment. Yet it is often the most cost-effective means to improve drying economy and may well be the most overlooked option. Some large spray-drying plants could easily afford to hire a technician full-time just to check the dryer feed, trigger immediate corrective action, and avoid unnecessary dilution.

A typical dryer's two biggest heat requirements are its evaporative load and the exhaust heat loss. Reducing the evaporation cuts demand about equally for the exhaust loss. It has no effect on the outlet air humidity. The feed moisture content should be reduced to the extent permitted by the feeder type and the characteristics of the feed and product. When they are practical, filtration, centrifuging, and other nonthermal separation processes, as well as evaporation, nearly always have lower overall costs than drying. None are able to reach the low product moistures that drying provides, but they are often cost-effective for predrying.

The required heat and the resulting production rate are proportional to the evaporation rate. An indicator of the change in heat and productivity for conventional drying processes is the ratio of evaporation E_v to production P_r. Inverting Eq. (7.2) puts the relation on a basis of unit weight of product,

$$\frac{E_v}{P_r} = \frac{100 - F_s - P_m}{F_s} \qquad (8.1)$$

where F_s is the feed solids and P_m is the product moisture, both in percent.

When the solids content changes, the required airflow by weight and volume also changes in about the same proportion as does E_v/P_r. The heat and the productivity change about equally, modified by changes in heat added by the feed components. For example, if a 30 percent solids feed is increased to 32 percent, the improvements in productivity and heat use are about 9 percent each, and in power use 12 to 15 percent. The air outlet humidity is unaffected.

If feed solids contents of 10, 50, and 90 percent are raised to 12, 52, and 92 percent, respectively, the required evaporation per unit weight of product decreases by 19, 8, and 22 percent. This assumes zero product moisture, which has little effect except at high feed solids content. As noted in Sec. 8.5.2, the main benefit of higher product moisture is usually the lower outlet air temperature it permits.

8.5.4   Preheat the feed

Feeds for directly heated dryers range from damp, granular substances to low-viscosity solutions. For solutions, emulsions, slurries, and other liquids it pays to preheat the feed as high as the product and process will allow. Heating solid feeds is more difficult, but the benefits are generally even greater on a percentage basis.

Calculations show that heating a 30 percent solids feed 100°F (56°C) will benefit heat, power, and productivity in a range of 9 to 15 percent, dropping off slightly at higher air temperatures. The savings assume little or no heating cost—the dryer exhaust can sometimes be used to preheat the feed. The outlet air humidity is increased moderately as the feed temperature increases.

For spray drying, when a wet scrubber is used as a backup collector, it can also serve as a feed preheater-evaporator. The feed must be below its maximum concentration and not have any adverse effect on scrubber operation, such as foaming or solids buildup. The benefit is double, reducing the dryer's evaporative load and adding heat by means other than in the airstream. But scrubbers cool the air to its saturation temperature, nullifying recovery of heat from the exhaust.

Most liquid feed properties that are relevant to drying are temperature dependent, and some, such as viscosity, are critical to dispersing the feed. In addition, they may affect attaining the desired product specifications. Raising the temperature of solutions usually improves atomization. Specifically, it may help to prevent filamentation of viscous feeds.

8.5.5   Reduce air in-leakage

Typically, a dryer operates at a slight negative pressure, so that any leakage is inward, avoiding an escape of powder, but generally concealing it. Eliminating a leak of only 5 percent of total airflow substantially reduces energy and raises productivity, but the outlet air humidity rises slightly. Leaks of 20 percent or more are common in older dryers. Unfortunately it is difficult to build and keep dryers airtight, especially large ones.

Certain high-temperature dryers have deliberate leaks to cool critical system parts, such as the motors used with some centrifugal atomizers. Other designs inject ambient or conditioned air into the drying zone to cool dried product before discharge. For the sake of energy savings and productivity, if humidity problems are not created, such flows should be minimized. Unfortunately the elimination of all leaks in some old dryers can cause operating difficulties. Some products are adversely affected by the increased air humidity—caused by the higher feed rate—unless the outlet air temperature is raised.

8.5.6   Preheat the supply air with the exhaust

About 20 to 40 percent of a direct dryer's heat goes out the exhaust; in some cases the loss can be over 60 percent. A portion of it can be reclaimed using an exhaust-to-supply air heat exchanger installed in the ducts. Heat is taken from the air leaving the collector and transferred to the supply air before it enters the heater.

Various designs are in common use, employing the exhaust to preheat the supply air or for space heating or other process needs. Up to about 30 percent of the total dryer heat load can be recovered in low-temperature drying systems. In high-temperature dryers the percentage is as low as 12 percent, but exchangers can still be justified because the total heat load—and the heat saved—is greater. Productivity remains unchanged, however, because there is no reduction in the amount of air needed to supply the heat for drying.

There can be three additional, somewhat unrealized, benefits in using these external heat exchangers.

1. If the drying system has a direct-fired heater, the heat saving by use of the exchanger reduces the amount of fuel burned. This also reduces the moisture formed by combustion, lowering the dryer's outlet humidity slightly.

2. The reduced formation of water vapor when using a direct-fired

heater also increases heat recovery above the exchanger's heat load. The added saving is explained in Sec. 6.4 and shown in Fig. 6.2.

3. Some additional heat will be available for recovery if the exchanger is in its usual position downstream of the system's blower. The heat added to the exhaust by compression and friction of the blower would be lost if there were no exchanger. The air temperature is raised about 0.5°F/in WG (1.1°C/kPa).

The heat recovery advantage of heat exchangers is slightly offset by the extra power cost from the added pressure drop. For typical installations this averages 2 to 8 in WG (0.5 to 2.0 kPa). An extra 4.0 in WG (1.0 kPa) was used in the examples, which penalizes total savings by only 6 to 12 percent. Net savings for less fuel but added power are 18 to 24 percent of total energy costs.

The most common types of exchangers used in drying system ducts are air to air and air to liquid. (The latter are sometimes called run-around or liquid-coupled.) Rotary recuperators and heat pipes are used less often.

Air-to-air units are bulky, and in existing installations sometimes cannot be fit into the space available, but they recapture the maximum heat. Figure 8.3 shows the percent of a dryer's heat load they can recover. The chart assumes that supply air and feed to the dryer are both at 60°F (16°C) and that the exhaust is cooled to its dew point.

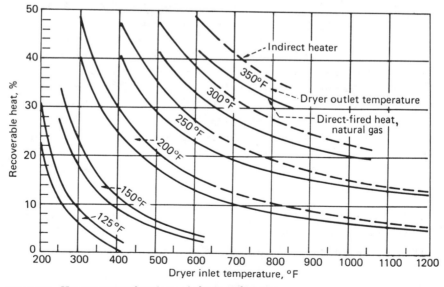

Figure 8.3 Heat recovery by air-to-air heat exchanger.

When a dryer uses indirect heating, the air temperature has a practical limit, usually 600 to 800°F (316 to 427°C). The dashed lines in Fig. 8.3 begin at the first part of that range. As noted earlier and indicated on the chart, any combustion moisture added to the air affects recovery. Benefits are greatest with indirect heating, least when firing with natural gas, and oil gives results between them.

Air-to-liquid recovery requires two exchangers, one in the supply duct, the other in the exhaust duct. Water or antifreeze is pumped between them to pick up heat from the exhaust and deliver it to the supply air. Less heat can be recovered than in the air-to-air type, but these units can be fit more easily into existing systems, requiring little or no duct rearranging. Tubing is commonly finned, thus helping to offset the poor heat transfer coefficient for air.

In any of the designs, if the exhaust is cooled below its dew point, additional heat can be recovered. The added saving can be greater than the cost of increasing the size of the heat exchanger. At higher exhaust temperatures, a point is reached where that relationship is reversed, and condensation becomes less cost-effective. But even in the range of greater savings, allowing condensation to occur in standard designs may cause trouble. It often exposes the units to acid gas corrosion, freezing, and buildup of escaped product.

Wash-water devices are commonly added to exchangers to flush off product buildup intermittently. Because condensation can aggravate this buildup, cooling is often limited to some point above the condensing temperature. In practice, however, it is difficult to avoid some condensation.

Minor product accumulation on the air side of exchanger surfaces has little effect on energy recovery, because the high heat transfer resistance for air governs the overall rate. Any sizable deposit of product would, of course, seriously affect performance and, eventually, pressure drop and airflow. Designs are available, at higher cost, that use glass or polished tubes, or tubes coated with fluorocarbon polymer. Most deposits that do form on these tubes are readily removed by a water flush.

8.5.7 Use two-stage drying

Removing the last fraction of moisture from the product requires a large portion of the exposure time and vessel volume, because the temperature driving force is diminishing. Two-stage drying overcomes this restraint by using a second dryer for final moisture removal. The product is dried to as high a moisture content as practical in the main dryer. It is then brought to its final condition in the secondary dryer, which is operated at more appropriate conditions. The technique is easiest to justify at high capacities, and it results in substantial improvement in both energy use and productivity.

Virtually all large powdered-milk dryers built recently have two stages. The first is a spray dryer that brings the powder to 6 to 7 percent moisture, with outlet air 20°F (11°C) cooler than for a single unit. The second is a comparatively small fluid-bed dryer that drops the moisture to the specified 3 percent. Overall energy savings and production increases are about 18 percent each, compared with a single-stage system. Also, the higher moisture content out of the first stage makes it possible to produce in the second stage a more desirable, *instantized* product (one that redisperses quickly).

The relatively low outlet temperature in the first stage can result in a higher-quality product. Alternatively, the same quality can be maintained and further cost and productivity benefits gained through other operating adjustments. These include higher inlet air temperature. For milk drying, the gain in productivity can total 40 percent; the benefit in energy savings is only slightly less.

Two-stage drying is shown in column 7 of Table 8.3. The first three rows list the air outlet temperature and the product moisture from the first stage. The benefits are 14 to 25 percent energy savings and 17 to 26 percent productivity gains. The second stage has offsetting energy costs that are not deducted, but they are relatively minor. In fact, the second-stage exhaust can sometimes serve as a partial heat source for the first stage.

During a test program in a laboratory, if the drying curve prepared shows the need for fast evaporation, followed by a long exposure time to reach low product moisture, two-stage drying is often the best arrangement. Some materials cannot be dried in a single stage. They may need a high heat transfer rate followed by a long period of moisture removal by diffusion. Or part of the drying operation may need different dryer types to handle sticky or other difficult conditions. In commercial installations both direct and indirect dryers have been used for both stages.

8.5.8 Use an internal heat exchanger

One or more heat exchangers, often referred to as coils, can be located inside certain dryers. The heat added is independent of the airstream, allowing the airflow rate to be reduced. Fluid-bed dryers are the principal type using these coils, which can be of tubular or plate design. Energy costs and productivity can be improved by one-third or more. Because of their limited use no data are given for them in the tables, but they are shown in Figs. 8.1 and 8.2.

Those two diagrams show that energy and productivity benefits for adding an internal exchanger are extremely high. The example assumes heat added by the exchanger of only 25 percent of the original dryer heat load, which is conservative. Some fluid-bed dryer coils approach a 90 percent contribution, resulting in a great reduction in airflow. The amount of surface may be limited or even eliminated, however, if higher airflow is needed for fluidization or to reduce air humidity. Fouling of exchanger surfaces precludes their use with sticky materials, at least until the sticky phase has ended.

Benefits only apply to the heating of the solids in the dryer. It does not apply to any heat transmitted to the air, which to some extent is unavoidable. Heating only the air inside a dryer, as commonly practiced in conveyor and tray dryers, provides no more advantage than a conventional external air heater.

8.5.9 Recycle some of the exhaust air

This option is used on large batch dryers and some others, such as rotary-tray units, for duties that can tolerate increased air humidity. When used for direct dryers, it is often for purposes other than drying efficiency, even though energy savings can be high. As described in Sec. 2.4, recycle systems are useful when a low oxygen level or reduced exhaust flow is needed. The productivity benefit is very small, and in some cases operating problems are difficult to overcome.

Data for recycling exhaust are not given in the tables, but Fig. 8.1 shows that the heat savings are close to, and parallel to, those for reclaiming heat from the exhaust with an external heat exchanger. This is to be expected because the operations are similar.

8.5.10 Change the heat source

The combustion in direct-fired heaters adds water vapor to the air. Of the common fuels, natural gas adds the most moisture, oil about 43 percent less, and indirect heating none.

If a dryer is switched from natural gas to fuel oil, the reduced amount of water vapor formed lowers the total heat requirement by about 4 percent. This change also lowers the outlet air humidity somewhat. Productivity and power use are not affected.

8.5.11 Insulate the drying zone

The heat loss from the drying vessel depends mainly on its surface area, its temperature, and the amount of insulation. The temperature is a function of the difference in temperatures between the air inside and outside the dryer. With cocurrent flow of air and product, the hot inlet air gives up its heat rapidly in causing evaporation, so most of the dryer's interior skin is contacted by temperatures approaching the cooler outlet temperature. Using 2.0 in of insulation, the heat loss is typically 0.25 to 0.50 percent of the heat load in dryers having large surface areas. Dryers with less surface, lower temperatures, and more insulation have even less heat loss. Thus heat loss is seldom an important influence on efficiency for dryers that are insulated.

If a fluid-bed or other large chamber dryer is not insulated, the heat loss is about 4 to 6 percent of the total heat load. On many dryers, particularly small ones, insulation is needed mainly to protect personnel from burns. Installing mineral insulation drops the outlet humidity a little and benefits energy and productivity 5 to 20 percent.

There may be a substantial heat loss from the hot inlet air duct if it is uninsulated. This would lower the air temperature entering the drying zone—and increase energy use—but it would not otherwise affect drying.

8.5.12   Some other options

There are various other ways to reduce or recover heat from drying operations.

1. Using a dryer's exhaust heat for other process heating or space heating, as noted before. Scheduling can be a problem, and backup systems are usually needed.

2. Also mentioned previously was the effect of higher product moisture. It has an increasing effect as feed solids increase.

3. A steam heater can be an economical preheater to offset the high cost of heating with electricity.

4. Flue gas or other inexpensive hot exhaust can serve as a heat source, but would be ruled out if its moisture content is too high. One possible alternative is to transfer some of the waste heat into a dryer's supply air with a heat exchanger. Another application is calcining, which needs predrying; both operations are usually handled together in a kiln. Flash drying the feed, with the kiln exhaust as the heat source, uses both energy and the kiln more efficiently.

5. Changing from water to an organic liquid may be worth considering. Most common organic solvents have latent heats of vaporization that are less than 20 percent that of water; alcohols are less than 38 percent. Chapter 13 describes some of their more important aspects for use in drying.

6. Another logical technique, but with limited applications, is the use of superheated solvent vapor as the drying gas, especially when

oxidation of the product must be avoided. Studies have shown steam to be a good candidate.

8.6 Overall Results

For each of the options in the examples the changes selected are conservative for those circumstances where changes are possible and practical. The inlet air temperature, for example, can often be raised more than 20°F (11°C) and still maintain good operating conditions and necessary product quality. Leakage can often be cut more than 5 percent, especially in old dryers.

The modest rise in inlet air temperature, together with the leakage reduction result in the least savings of the options in Table 8.3. All the other options would have greater benefits. But two or more options are not additive. The correct total requires a calculation combining the options.

Although the main advantages of dryer optimization are immediate economic gains, a real, if less tangible, benefit is to the environment. Cutting back on the dryers' enormous consumption of fossil fuels helps to defer their depletion and reduce the effects of carbon dioxide on the atmosphere.

Selecting by Testing

9.1 Introduction

Testing is the key to success for both dryer users and dryer manufacturers. Designing a dryer without dependable trials tends to bring unwelcome results and is almost never done. On one of those rare occasions, a large dryer firm, having some experience with the product, guaranteed the performance of its dryer without running any tests. Incredibly, the firm agreed to pay for all off-grade product from its dryer. The failure of the dryer was no surprise to industry insiders. The loss to the manufacturer was considerable, but the buyer also suffered in lost production and wasted use of resources.

The many available designs illustrate that no one type of dryer is suitable for all, or even a majority of duties. Some dryers have niches, established by tradition or by their known superiority on certain classes of products. It is unusual, however, for only one type to be both acceptable and cost-effective. Thus a search is needed to find the right dryer, especially for an untested material. Because of the range of materials and dryer types, only a brief description can be given here of the steps to find the right test facility and the best dryer, and to illustrate how tests are run for the common designs.

9.2 Why Test?

Most process equipment can be calculated from general, empirical equations that relate the design to material flow rates and properties. The basic relations are established by testing; further tests are not usually needed. The lack of such relations for dryers means that virtually every application requires testing to answer these questions:

1. Is drying feasible?
2. Does the dried product have the desired properties?
3. Does the drying system run without buildup or other equipment problems?
4. What are the best operating conditions?

Finding the answers can be arduous, and may involve hunting for acceptable conditions for each of several dryer types. But only when testing is completed can costs be figured so that different systems can be compared.

Some materials have more complex test procedures than others. Much depends on the action in the dryer—solids resting in static beds need less, agitated beds need more, and droplets sprayed into a fast-moving stream of air need the most.

9.3 Selecting Test Laboratories

Dryer selection begins with a search for good test laboratories. No dryer manufacturer has the ideal—a full range of direct and indirect dryers—so the best procedure is to endure the cost and inconvenience of investigating and running tests at several laboratories.

Some applications have special needs. For example, separate facilities can provide greater safety or cleanliness; large test dryers assure more accurate scaleup; test units set up where the feed originates may be needed for better security, control of toxicity, or other reasons. But the main goal of testing is to determine suitability and to generate design data for scaling up to commercial systems.

Inspecting. Visiting a manufacturer's test laboratory will show its general state of cleanliness and orderliness; dust and vapor controls are clues. It will also show whether the equipment, techniques, and instruments are appropriate and up to date. Safe practices should be observed and necessary precautions taken to prevent fires and explosions. All parts of the test equipment that contact feed and product should be corrosion-resistant.

Before committing to tests it is also advisable to consider the capability of personnel and equipment, and the types of dryers and their versatility. Equipment operating features are described in Sec. 9.7. Any special facilities that will be used should be inspected—auxiliary equipment for preforming feeds or modifying products, separate facilities for sanitary operation, and drying systems for making large quantities.

Personnel. The various levels of personnel at test laboratories are an indication of the value placed on testing. In addition to the manager and operator, larger firms may employ all of the following:

Technician. Records data and assists the operator

Applications engineer. Knows the product and its processing requirements; prepares estimates and quotations

Laboratory engineer. Understands the equipment and its limitations; supervises tests; works with the client's representatives; writes the reports

In smaller firms it is more likely that the operator records the data, technicians not involved in testing prepare quotations, and the test manager performs the combined duties of applications and laboratory engineers. In this arrangement, a sales engineer with an understanding of the product and process usually is the customer contact. In very small firms it is not unusual for a single engineer to handle most of the technical aspects of testing, plus the sales, quotation, project engineering, and any field startup duties as well. This policy has both benefits and limitations.

Equipment versatility. The major elements of versatility include good procedures, capable operators, and test units that can simulate each of the company's dryer designs and can be modified to approximate their essential features. These modifications or adjustments are different on every type of dryer.

The significance of versatility is seen in testing an unknown material considered to be a candidate for disc or paddle drying. A series of tests in a moisture balance can be run to give some idea of the required exposure time. If not too long, a high-speed paddle dryer is indicated; otherwise the disc dryer would probably be better. Very long exposure times need the low-speed paddle dryer, but if the feed is very wet, a two-stage operation may be called for. In cases like this there are advantages in having a stable of dryers to suit a range of conditions.

Although laboratories can usually prepare and modify feeds and get the most out of their dryers, they can seldom determine the most cost-effective overall system. For example, a slurry might be sprayed into a dryer, or it might first be filtered or centrifuged, then fed as a wet solid (if it has acceptable characteristics) to another dryer, possibly at much lower overall cost. Test personnel often can advise on such projects, but they are not likely to have the equipment and expertise necessary to make full analyses.

Separate facilities for some products. To meet strict standards of purity for testing foods, pharmaceuticals, and some other products, the accepted practice is to isolate the system so the equipment, floors, and all else can be kept contamination-free. Clients often send inspectors, who take swab samples, and sometimes have preliminary tests run on a simulant and then check the product for purity. Standards are especially strict for contract drying (custom drying) of foods and other ingestibles.

Dangerous materials. Dryer testing laboratories are not equipped to handle very toxic or otherwise dangerous substances. When it is essential to get data, tests are run on a simulant that dries like the active substance. An example is avoiding the dangers of beryllia by substituting aluminum oxide for both testing and starting up commercial units.

A more insidious danger is posed by substances, such as lead compounds, that can accumulate in a worker's system over time and be difficult to treat. Workers exposed to such materials should be examined on a regular basis and kept from contact with these substances if their condition passes a critical threshold. In addition, there should be no release or improper disposal of harmful substances.

Testing outside the laboratory. Operating data for designing commercial units must come from either laboratory tests or actual plant trials. Depending on the application, the equipment, the instruments, and how well the tests are run, it may be possible to dispense with conventional tests either partly or completely. Whether run in the user's plant or at a contract drying firm, the tests should be witnessed by personnel with testing experience. The same care and procedures as in the laboratory are needed if the results are to be acceptable.

Installing a test unit at the feed source solves most problems of feed transport and degradation, helps control safety, eases time pressures, simplifies maintaining confidentiality, and gives employees easier access to the trials. For long-term projects this reduces overall costs. Needed expertise can be contracted from the manufacturer or an independent consultant.

It is common practice for a user to buy a new or used test dryer for a material that is hazardous or degrades quickly, or if trials might corrode or otherwise damage the equipment. An alternative is to rent a dryer, unless the substance is radioactive or in some other way would impair the equipment. Rentals, which are offered by some larger used equipment dealers as well as manufacturers, are in most cases easy to set up. Indirect test dryers may not need product collectors and are relatively easy to operate. The simplest are free-standing and complete; they are just bolted down and connected to utilities. The usual arrangements are for a minimum of 3 months, with the user paying for transportation, cleaning, and any wear or damage.

Contract drying.   Before installation of a newly purchased drying system, bulk quantities of the dried product are often needed to investigate further processing or to survey markets. Most manufacturers' equipment is too small to make the required quantities and meet strict product specifications, so clients turn to contract drying firms. Occasionally the ultimate customer for the dried product, after examining the newly dried material, will no longer accept the previous material. Contract drying is then especially important to both parties.

Test fees.   A test laboratory is a sizable investment in equipment, building, instruments, and facilities, plus the expenses of operating, maintaining, and cleaning. Recouping some of these costs by charging test fees is widely accepted, although the fees are sometimes waived for sales reasons. Some firms, especially those outside the United States, seldom or never charge fees. They increase equipment prices instead. But sharing costs among all clients seems more equitable than charging only those who purchase new dryers.

Tests are sometimes scheduled merely to obtain dry material for use in further research, or to get data for building dryers elsewhere. Charging fees helps the laboratory to cut down on unpromising exploratory work, which might otherwise become burdensome. Manufacturers insist that their test operations are not profitable, and that fees do not even cover total operating costs. But fees may be set more by tradition and business conditions than by actual costs. They vary widely and are often two to four times full labor costs.

9.4   Overview of Testing

Even after having decided where to test, narrowing the choice of dryer can be a long process. During trials, changes may have to be made in the type of dryer or its features, in the feed, or even in product expectations. The feed's characteristics, the ease in feeding it, and the dryer's ability to accept it have major effects on the suitability of a specific design. Also important are product specifications and proposed production rate, which may rule out designs unable to do the job in a single unit.

Outlining goals.   Before trials begin, a questionnaire covering the important details about the material, especially hazards, should be completed. Past experience with similar materials is only a guide in setting up a test program. This involves a full discussion of objectives, procedures, the nature of the feed, any hazards, the disposal of all materials, and the product requirements. If the important product properties cannot be measured accurately between runs, an acceptable, quick method has to be arranged; otherwise test results might be meaningless. A sample test questionnaire is shown in Fig. 9.1.

If the ultimate objective is to produce quantities that are larger than test quantities, the trials must yield reliable data for scaling up

| COMPANY Name, Address, Individual | Date |
| | Telephone |

PROCESS Previous processing method; operating conditions

Status: ☐ Production ☐ Research ☐ Development ☐ Other

Objectives: ☐ Determine feasibility ☐ Obtain samples ☐ Equipment quotation ☐ Other

Other processing information

| MATERIAL Name | Chemical name or formula |

| FEED Solution, slurry, filter cake, other | % Solids | Temp. | Toxic? |

Softening: (   )°____   Melting: (   )°____   (Note whether temperature is °F or °C)

Stickiness: (   )°____   Others:(   ) (   )°____

| Specific Heat: | Max. Grain Size: (   ) ☐ Mesh ☐ Micron |
| Average Grain Size: (   ) ☐ Mesh ☐ Micron | Min. Grain Size: (   ) ☐ Mesh ☐ Micron |

| Rate lb/h | Sp. Gr. | Viscosity |
| | | Can viscosity be lowered by heating?   Max temp/time |

Feed preparation steps

Particle size in feed

PRODUCT Rate lb/h	REQUIRED PROPERTIES				
	% Prod. moist.	Particle size	Bulk density	Flammable?	Critical temp.
			Temperature °____	Explosive?	
Hygroscopic?	Heat stability Temp/time		Thermosetting or thermoplastic Crit. temp.		Ht. of cryst'n (or sol'n)

Product testing methods:

Toxic? Supply complete toxicity information.

Figure 9.1 Test questionnaire.

to a commercial unit. Discussions should cover the project's goals. Will equipment be purchased, or is the product to be dried on existing equipment or by contract drying? How much feed will be needed for testing and what is the projected production rate? The timing of the project is important—how soon are estimates or quotations needed, and will experimental design or other time-consuming techniques be used? What data and product samples are available from other tests or from production? From this information and the test questionnaire, a program can be drawn up, although it may have to be altered as work progresses.

Cost, cost analysis, and optimizing. The cost of a test program is only a small fraction of a project's total expenditures for engineering, equipment, installation, and facilities. But inferior test work will almost certainly incur unknown, and higher, costs from cutting short a promising project; or from selecting a less profitable dryer.

The ultimate operating cost is ignored in the early stages of most projects; in later stages it usually dominates. A project once begun, however, has an inertia that inhibits changes and receptivity to new ideas. To permit making changes in a timely manner, the total operating cost of the proposed system should be estimated during trials. Otherwise successful drying may allow continuing a project that should have been canceled earlier.

Calculations given in Chap. 8 show how various process options reduce the costs of heat, power, and equipment. Because most of the steps relate to reducing the airflow, indirect dryers benefit less in absolute values, but for those options that apply, they benefit more as a percent of the totals.

A prime example of optimization is two-stage drying. Certain direct dryer applications can be improved by 20 to 25 percent in both energy use and dryer size, the main components of the total operating cost. The cost benefits or penalties of these options can be analyzed on a spreadsheet during test programs to help find the best conditions. Estimated values can be used for proposed operating rates and utility costs. But for reasonable accuracy the analyst should have the capability to compute dryer airflows and heat loads from the operating data. Experimental design procedures can be made a part of this work.

9.5 Feeds and Feeding Problems

The amount of feed material needed for testing can be approximated for solid feeds from the estimated heat transfer rate U, the heating surface A, and the temperature difference T_d, which is the heat source temperature minus 212°F (100°C). The heat load for indirect dryers is roughly 1000 (1500 for direct dryers) times the evaporation rate E_v, lb/h (kg/h);

$$1000E_v = UAT_d, \qquad E_v = \frac{UAT_d}{1000} \qquad (9.1)$$

and
$$F_d = E_v\left(1 + \frac{100 - F_m}{F_m - P_m}\right) \qquad (9.2)$$

where F_m and P_m are feed and product moistures and F_d is the feed needed for 1 h of testing. The amount required for the test program can then be figured from the estimated testing time: the number of tests times the exposure time normal for such a material.

Problems from shipping.   Feed should be checked on arrival to be sure it is in the condition intended for drying. During shipment from the plant it may have changed, perhaps irreversibly. A liquid or wet solid that has frozen, for example, even when thawed, may differ from the original. A liquid feed may also have settled, crystallized, coagulated, reacted, or changed in some other way. Sometimes the drum lining dissolves or partially separates from the metal and joins the feed. Restoring a feed to its right condition may be a challenge, exceeded only by its arriving too late for the scheduled test.

To aid in examining, removing, and restoring, all feeds, liquid or solid, should be sent in open-top containers. And liquids should be completely homogenized by thorough mixing before, and kept well-mixed during the runs.

Treating liquids.   Diluting or heating liquids to improve spraying is sometimes necessary in small-scale tests, but should only be done as a last resort if it is not to be done on the eventual unit. (That also applies to adjusting the pH and using additives, such as phosphates or deflocculants, intended to improve drying characteristics.) Diluting increases both energy use and equipment size. Conversely, heating decreases both, but it may alter the material.

Tackiness.   Some solids go through a tacky (or sticky) phase while drying, and this rules out a drying method in which material sticks to surfaces and continues to build up. Even in a spray dryer, when airborne particles first contact the chamber or ducts, they must be past this tacky phase. Similar problems are a low melting point and the release of molecularly bound water that may limit metal and product temperatures.

As backmix for tacky feeds or as a starting bed, or both, some dry product may be needed. A sample of dry material, in its desired final form, is often useful to compare against the product. (In addition, a suitable solvent may be essential for cleaning the product or its fused or burned derivatives from equipment walls.)

Substantial backmixing increases the required vessel size. This is modified to some extent, however, by the temperature of the recycled product. Adding warm material directly into the feed reduces the need for the slower heating by the hot surface. Nevertheless, backmixing reduces overall drying efficiency and should be minimized.

Preliminary examination.   Liquids should be checked for properties that might affect feeding or drying. A slurry that settles quickly may require special agitation. Any particles large enough to plug the smallest opening in the feed delivery system must be screened out—a 30-mesh screen is used for feeds that are sprayed. If a paste is thixotropic, stirring it may make it thin enough for spraying. Complex feeds may have to be modified—to finer grit solids in a ceramic slip, for instance. Feed viscosity has a significant effect on pumping and spraying, thus on product characteristics. It can be measured on a viscometer and adjusted by changing temperature or concentration, if permitted.

Moisture balances are used to dry both liquid and solid samples. They may require several trials to find the settings that avoid any burning or surface hardening and give consistent readings. Plotting the moisture content readings against time roughly simulates the drying process and helps predict the exposure time. Moisture balance readings are only approximate, but the differences between readings usually give indications that are close enough for test work. They can also help to estimate tackiness, the softening or melting point, the film-forming tendency, heat sensitivity, and other characteristics.

Except for liquid-fed dryers, most can easily accept granular or crumbly solids. A crude but useful check of acceptable consistency is to knead a handful of feed in a small plastic bag. Another quick test is to form a lump of solid feed. If it breaks up when dropped on a hard surface, it probably can be fed to a dryer that actively stirs the material. On the other hand, if it does not break up, or if it sticks when thrown against a wall, it may stick inside the dryer, especially if surfaces are hot. Such a material has to be backmixed with dry product before drying, or fed into a bed of dry material.

An instrument called a hot bench—a kind of graduated hot plate— is commonly used to make *stick* tests on feeds. It can be calibrated with substances of known drying properties, and it indicates where a solid or liquid sample melts or becomes sticky, thus defining allowable product and metal temperatures. It may also indicate if a bed of dry material will be needed to start the unit.

Feeding methods. Selection of a liquid feeding method is influenced by consistency. Low-viscosity liquids, for example, can be fed to a small spray dryer, at the necessary uniform flow, using a graduated glass cylinder with petcock, which is easy to set up and lets the feed be seen and stirred. For materials that must be pumped, the peristaltic (roller-tube) pump can be used on many feeds. Gear pumps are satisfactory if the feed is not abrasive. Progressive cavity pumps give a smooth, positive flow, and are preferred for providing the shear forces that make thixotropic pastes sprayable.

For solids a variety of feeders are available; some applications need a variable-speed drive. The volumetric screw feeder is favored, but others used are vibrating tray, vibrating hopper, rotating table, and venturi aspirators (for flash dryers). Each has its advocates, but because of setup times, and possibly the need of a hopper, careful hand feeding with a scoop is sometimes sufficiently uniform and more practical.

To minimize air leakage with solid feeds, the feed entry can be at least partly choked off by the feed itself. Preferred for this duty are special screw feeders, double flapper valve, and rotary air lock.

9.6 Product Properties

For most products, the important properties that are affected by operating conditions have to be determined between runs and compared with the specifications. Evaluating methods vary widely; one fine chemical firm has three separate devices for measuring the bulk density of four related products. Flow characteristics, such as angle of repose, are often important in subsequent use, or even in the drying operation itself, as in rotary dryers and rotating tray dryers.

Particle size. Size analysis is vital when particles are formed or affected by the drying process. Attrition in ducts and cyclones is apt to have an effect, especially on fragile particles. The proportion of various size ranges can be determined by sieve analysis on a set of vibrated screens, by sedimentation analysis, or by instruments that automatically analyze distribution by a light-scattering technique. The latter are fast enough to get data between runs, but if no data are needed, a visual check using a microscope with a sizing grid is often acceptable.

Moisture content. Final moisture content is almost always a condition of product acceptance, and thus a guide for setting the heating medium temperature or residence time for the next test run. The same moisture balance used to give the feed a preliminary check is used for moisture analysis. But samples of difficult products are returned to the client for analysis, which may take more than a day. Vacuum desiccators or ovens give the necessary accuracy, as do electrical resistance, chemical, and distillation methods, and they can also be used to roughly calibrate moisture balances for specific materials.

The final product from a dryer with short residence time may be dry only on the surface. As the internal moisture spreads, the surface may become wet and fuse the particles together. The product must be at or below the acceptable average moisture content; otherwise the entire mass in the container may fuse into a single block. A similar problem can be caused by temperature equilibrating in a product that has a low melting point component.

9.7 Drying Tests

Testing is a sequence of go or no-go trials, each requiring success before starting the next. The steps vary with the application, but in general, after examining the feed's handling properties, the type of dryer, feeder, and initial operating conditions are selected. Then the feeder's ability to deliver the material is tried, and if necessary it is adjusted or changed—or the feed is modified. When that test is completed, actual drying runs can begin, adjusting conditions and equipment options if necessary, until the product is in the desired form and the equipment surfaces are relatively clean.

After each run the equipment walls and ducts are examined. The amount of any recovered buildup is weighed and totaled with the regular collection. Buildup has to be accounted for. If it accumulates, continuous operation in a commercial unit may not be possible.

Test laboratories are under pressure to run the most tests in the least time. There is little opportunity for refinements, such as reruns to check replication of data or to optimize the operating conditions. A useful addition to normal direct dryer test routines, however, is to measure the dew-point or wet-bulb temperatures at the dryer outlet. Knowing one of these may save time in the long run by enabling the outlet condition to be verified quickly on a psychrometric chart. Otherwise some condition such as heat loss, heat of crystallization, or a leak could move the outlet position unexpectedly and nullify the test objectives.

For most testing, but not all, the size of a dryer affects the results, and failure in a small unit does not necessarily rule out success in a larger one. Bench-scale dryers, for instance, at the lowest end of the scale, are limited to easy-to-dry materials that are costly or for other reasons must be accurately accounted for. But for some materials they can check feasibility quickly and give approximate exposure time, temperature, and drying results. Designs made of glass give visibility and corrosion resistance not available in conventional units.

As in the previous chapters, dryer listings in the following sections are alphabetical within the indirect- and direct-heated divisions, using names either in common usage or considered acceptably descriptive.

9.7.1 Indirect dryers

Except for temperature and feed rate, most operating conditions can only be adjusted between runs. These changes help to find the best conditions and to correct problems, such as any undesirable shrinking or agglomerating. Testing for the best conditions is often made difficult by the following two relationships that are important in indirect dryers.

1. Increased agitation raises the heat transfer rate (until serious fluidization occurs). But it may lower the exposure time and may also increase short-circuiting of particles to the outlet, which can increase nonuniformity between particles.

2. A wet feed requires a high evaporation rate. Thus it is best dried at a high heat transfer rate, but exposure time can be relatively short. Conversely, with bound moisture to be removed, a long exposure time is needed. To remove large amounts of both free and bound moisture, often the best answer is two drying stages.

The data needed for scaleup set the direction of a test program. For heat transfer dominated drying, information is needed for calculating the heat transfer rate. When the drying mode is diffusion dominated, the mass flow rate and exposure time are the critical factors. For all of the bed-type dryers, and especially for the rotaries, it is important to know the *fill*, or loading ratio, throughout the dryer. This is the ratio of the bed volume over the vessel volume, and it is affected by the factors governing solids transport and by shrinkage.

A general condition influencing many applications are plastic and shearing phases that occur before the dry granular phases are reached. The peak power required may be three times that at the beginning or end of the cycle. For this reason it is not possible to extrapolate power needs from a test run on either wet feed or dry product. Power reaches its maximum at some point between.

Exposure time in the indirect dryers covers a wide range. It is determined for bed-type dryers from the volumetric flow rate of material, divided into the volume of holdup, measured after shutting down. The variation in exposure times and the inability to predict the time-consuming problems of testing make it impossible to narrow the time to complete a test program any closer than a range of 2 to 10 days.

Disc dryers. A test disc dryer generally has a single shaft with closely spaced, flat discs rotating at a tip speed that can be varied from 30 to 240 ft/min. Both discs and shaft may be zoned for heating control. The jacket of the housing is also heated and may have up to six zones.

Disc dryers work best on feeds for which a drying time of 15 to 60 min is indicated on a moisture analyzer. Feeds should be free-flowing. If they stick to hot surfaces, backmixing with dry powder may be required. On most tests any air, or other gas used, is just a sweep to clear out the vapor. Sometimes large flows of heated gas are needed to partially fluidize the bed for better heat transfer, especially for products for which high moisture contents are acceptable. The heating surface per unit of volume is high; so are the heat transfer rates, but lower than in high-speed paddle dryers. The exposure time can be varied up to several hours, which is beneficial for the more common application of diffusion dominated drying.

Disc dryers with segmented discs. Called paddle dryers but resembling disc dryers, a test unit may have a 10-ft^3 volume and twin adjustable speed rotors. Purge airflow is usually very low, but has to be high enough to prevent condensation of moisture on uninsulated areas, which would drop back into the product. When aiming for very low product moisture, the air is heated to keep it well below saturation; it enters at both ends and exits at the center. The exposure time can be varied from about 10 min to 2 h. Trough jacket, discs, and rotors all are heated, but the rate of heat transfer through the discs is much higher because of the scraping action.

Drying can be either heat transfer or diffusion dominated. Some wet, even sticky, feeds can be handled because of the agitator action, but for the most difficult feeds backmixing before drying is needed. Control of the wide-ranging exposure time is by feed rate, weir height, agitator design and speed, and slope of the vessel.

Paddle dryers—high speed. This type can operate as either an indirect or a direct unit, and testing is in two separate units. The first stage may be an 8-in-diameter horizontal tube 4 ft long, with about 8 ft^2 of heat transfer surface. The unheated agitator typically rotates at about 2000-ft/min tip speed, close to the cylindrical jacketed housing. Some heated air is used for most applications. The small size exaggerates edge effects and carryover of solids in the air, limiting the ability to scale up accurately. So second-stage runs are made in a unit with about five times the heating surface in multiple zones.

The heat transfer rate is high, making it well-suited to drying that is dominated by heat transfer. The exposure time is generally less than 10 to 15 min. When diffusion drying is also needed for reaching low moisture contents, a secondary unit is added in series, usually a low-speed paddle dryer.

A wide range of feed consistencies can be handled with proper adjustments to the agitator's clearance, angle settings, and speed. These adjustments also control the exposure time. Because of the intense action, wet or sticky feeds are less likely to need backmixing, but when necessary, it can often be done inside the unit.

Paddle dryers—low speed.   The low-speed paddle dryer is very different from the high-speed unit. Its T-bar agitators rotate in a horizontal, U-shaped or cylindrical shell. It is used for materials needing long exposure times to reach low moisture contents. Thus it often serves as a secondary dryer. The exposure time can be 1 h or more. Because diffusion governs, the main application is removing small amounts of firmly trapped moisture, usually at a low temperature. For such duties the low heat transfer rate and surface are not a disadvantage.

Agitator action moves the solids through the dryer and mixes the bed gently with little attrition. Procedures are similar to those of the high-speed paddle dryer, except that testing is in a single stage.

Steam tube rotary dryers.   These indirect units are heated by steam tubes in a cylindrical shell, and a test unit typically has a 12-in diameter and is 8 ft long. Before it enters, the feed must be free-flowing and often needs prior backmixing to avoid buildup on the tubes. Types of feeder include mixer (for backmixing), screw conveyor (for sealing against in-leakage of air), and vibratory tray. Sometimes a small sweep of air is used.

The tubes cut into the bed of material to mix and to give a slight showering action. The shell rotation speed can be varied from 1 to 10 r/min. Dams and slope of the shell are adjustable to regulate the loading, which is in the range of about 15 to 20 percent of the shell volume. The exposure time (about 20 min to 2 h) and the loading are critical for successful results.

9.7.2   Direct dryers

Because the amount of airflow is the principal cost influence on direct dryers, testing aims to minimize it with these techniques.

1. Use the highest practical inlet air temperature and outlet air humidity, which is the lowest outlet air temperature.

2. Add heat by some other means, as by heating the feed. Indirect heating inside the dryer is very effective when practical.

3. Recover heat from the exhaust air.

On the other hand, a minimum airflow is needed in some applications to fluidize or convey the solids. In addition, a minimum may be imposed to keep the humidity low enough at the dryer outlet; otherwise moisture would condense at cool spots, resulting in a high-moisture product.

Details are given of spray dryer testing to show why scaleup can be an uncertain extrapolation, even with good data. This is particularly true when either the design or the size of the test chamber is much different from that of the ultimate unit. In addition, high production requirements entail a program of at least two, and sometimes three, sets of tests in successively larger units for scaling up to the final size with some confidence.

Conveyor dryers.   A test conveyor dryer is a small batch unit, with an air recirculator that allows independent control of airflow rate and direction. Normally the first phase of testing is to dry a sample of the feed overnight in a vacuum oven. Results indicate drying difficulties and how the test should be set up. The feed may have to be extruded and cut into small cylinders (noodles) or otherwise preformed. Feed is spread manually to a depth based on experience and on the oven test data, typically to 2 to 4 in (5 to 10 cm).

Operating conditions are made to simulate those in a continuous unit. Air is directed up or down through the static bed, and humidity is regulated by adding heat and injecting steam, if needed. Because exposure times vary widely—from 5 min to 4 h or longer—the number of tests per day range from 2 to 14. Thus a test program can take a day or as long as a week.

The static nature of the process allows conditions to be measured with no time lag. Automatic, continuous recording of product moisture gives data for a drying curve. Other observations include wet- and dry-bulb temperatures, shrinkage, and the amounts, if any, of fines generated and product buildup on equipment. Scaleup from batch to commercial-sized continuous operation is based largely on experience, backed by the test data which, compared to most other dryers, are relatively accurate.

Flash dryers.   A typical test flash dryer has a vertical tube, with a 10-in^2 (65-cm^2) round or square cross section, 20 to 30 ft (6 to 9 m) long, with venturi, screw conveyor, or other solids feeder. Sometimes a venturi aspirator follows a feeder to introduce the material in an air-stream. Second-stage tests may be needed for scaling up to high capacities, for which a unit of about 50 in^2 (323 cm^2) is typical.

Some systems have a classifier that will remove lighter particles (assumed to be dry) and recycle heavier ones (assumed to be still wet) to give them a longer drying time. Because the residence time is less than 1 s, feeds must be relatively granular and any lumps have to be milled. Backmixing is sometimes needed as well.

To some extent, however, small particles traveling at high speed collide with, and tend to break up, larger slower ones; fragile particles are readily shattered. Virtually all drying is in the heat transfer mode, because time is too short for diffusion drying.

Feed containing about 40 lb (18 kg) of water is enough for the usual test period of 1 day. But a full test program, if a second stage is needed, takes 3 to 5 days.

The feed inlet area and mill, even if the unit has inspection ports, are hard to examine and harder to clean without disassembling. Thus the highest inlet and lowest outlet temperatures are sought cautiously. The dispersing (or kicker) mill, with the airstream going through or bypassing it, breaks up wet agglomerates and gives them a vertical boost. The minimum air conveying velocity must be determined; it is 3000 to 5000 ft/min (15 to 25 m/s) for most materials.

Fluid-bed dryers.   Initial fluid-bed tests are run in a simple, vertical batch fluidization column with diameters as large as 12 in (30 cm) or as small as 3 in (8 cm). Data are taken of air velocity against both pressure drop and fines carryover to find the gas velocity range—from incipient fluidization to the limit of acceptable powder loss. Bed moistures and temperatures are recorded against drying time to establish a drying curve. The constant-rate zone is accomplished by backmixed design. The falling-rate zone is usually tested in baffled chamber designs to provide the needed exposure time with the least short-circuiting.

Feeds are most often wet solids, but spraying of liquid feeds is not uncommon. One system incorporates a spray dryer discharging into a fluid bed for agglomerating particles to large granules.

The batch column data may be used to scale up to commercial designs for all but extremely wet or otherwise difficult-to-fluidize materials. Another practice, however, is to follow up with runs in a continuous unit having a bed area of about 10 ft^2 (0.9 m^2) and adaptable to up to three zones. Weir heights in each zone are set from batch-test residence time data. The time required for runs and cleanup is normally about 4 days.

Rotary dryers—direct.   Direct-heated test rotary dryers are typically 2 ft (61 cm) in diameter and about 16 ft (5 m) long. This size allows for wide variations in residence times, from 1 min to 4 h. Airflow can be parallel or countercurrent, and the rate can be controlled. To determine the best operating parameters, the flights, dams, and the slope and rotation speed (1 to 10 r/min) of the cylinder are all adjustable. Typical feeders are the screw, vibratory, and mixer types. In addition, some liquids can be sprayed into the unit; sludges can sometimes be pumped in. Among the many applications are those that require self-inertizing the air.

On some applications equipment modifications are required, for which an on-site shop is essential to keep test programs within an average of 3 or 4 days. Some design elements of the ultimate commercial unit can be optimized from the test setup and results.

Spray dryers—first stage.   There are three basic designs for first-stage testing. All are typically 30 in (76 cm) in diameter, but cylinder heights are 2.5 to 8 ft (0.8 to 2.4 m). Some permit changing the type of atomization, airflows, and product outlets. Because of the small dryer size and high speed of the droplets, the particle size and other properties are restricted. First-stage testing occasionally is bypassed, but that often brings unpleasant surprises and costly cleanups.

Most first-stage trials are conducted in the *conical* dryer, which has a cylinder with the height equal to its diameter, mounted on a cone. Atomization is by spinning disc or nozzle. For longer residence times, or average particle sizes up to 80 to 100 μm, the *tower* or *tall-form* dryer is used, which has a cylinder two or three diameters high. For very heat-sensitive materials, the best design is the *flat-bottom* or *flat-base* dryer, which is similar to the conical unit except for a flat or dished base instead of the cone. It has an air-introducing sweeper or air broom to remove and, in some variations, to cool the product.

The feed-airflow pattern is cocurrent in all these designs, but countercurrent and mixed airflow units and other designs are available for more specialized duties. Usually 5 gal (19 L) of feed are sufficient for the typical 1 day of testing (up to 12 runs). All these units can produce, quite faithfully, some properties, but not large particles, and air temperatures are only rough estimates.

Spray dryers—other stages.   Second-stage test dryers are usually 6 to 7 ft (2 m) in diameter and can make particles two or three times larger than those of the first-stage dryers. The minimum program, including cleanup, ranges from 2 to 10 days, but special problems take more time. Feed requirement is about 200 gal (760 L), and only about four to six runs (which includes the cleaning) can be made in two shifts. For jobs that are in some way special, for example, jobs that will operate at high production rates or make large particles, runs are made on a still larger unit. Its size may be chosen to suit the project, or it may be necessary to compromise on what is available. For many projects a 16-ft (5-m) diameter is acceptable.

Spray dryers—test details.   Temperatures are set conservatively at the start, with some guidance from a moisture balance check. In the conical test dryer the spray of feed travels only 14 in (36 cm) from a spinning disc and 30 in (76 cm) from a nozzle before reaching the chamber wall. Interior surfaces are examined after each run because most products cling to the walls if not dry.

The nature of any roof and wall buildup has to be identified—*sticky wet* (not dried sufficiently) or *sticky hot* (melted). Its degree and type, from loose powder to a fused layer, indicate what temperature adjustments have to be made. If the walls are clean, temperatures probably can be extended—inlet raised and outlet lowered. If the coating brushes off easily, conditions may be near the optimum for a unit of that size. If it has fused to the walls or the coating is heavy, the inlet air is too hot or the outlet too cool, or both. Or, possibly, the material cannot be spray-dried, at least in that design. Material stuck to the roof indicates that the inlet temperature is too high, and an uneven buildup on walls indicates an atomization problem.

In the second stage more accurate data are obtained and more is learned about pumping and handling the feed. In many cases the particle size, bulk density, and other properties of the commercial product can be obtained closely enough for reliable scaleup. If not, runs have to be made in a bigger unit.

9.7.3 Miscellaneous test work

Tests in larger equipment.   Large dryers of various types are needed for drying bulk lots, but most test facilities are not equipped for this work. Some types of big dryers are available at contract dryers, which often specialize in a single class of product such as general chemicals, ceramics, or foods and pharmaceuticals. It is unlikely, however, that a laboratory will be found to dry toxic or hazardous substances.

Cleaning.   Contamination of the next product to be tested is avoided by thorough cleanout after each program. Between runs on the same product cleaning can be less fussy, but should be thorough enough to minimize affecting the next run's inspection and product samples. A complete wash-down between runs would be time-consuming and is seldom necessary.

Product collection.   Even though many commercial installations use a bag filter collector, the best collection for most small test systems is by cyclone and wet scrubber. When a bag filter is mandatory, as when all the fines must be collected, clean bags are essential. But responsibilities can become clouded, as for the disposal of dusts from different dryers, or if cross contamination causes exothermic reactions and fires. Finally, a backup wet scrubber may be required anyway to contain bag leaks or noxious gases.

On large test systems bag collectors are more common. The bags should be identified and stored separately for each specific client, or even owned by them. Separated by product as well, they are cleaned each time the product is changed. Most contract drying facilities, however, use the cyclone–wet scrubber system for the least number of problems.

Test data and reports.   A prime objective is to get as much information as practical in one shift. Data taken on the feed include its type, percent moisture, specific gravity or density, and temperature when fed. Operating data vary some with the dryer, but usually include the temperature of the air before the heater, into and out of the dryer, type of heat source, feed device, airflow directions and rates, and any relevant equipment features. Compared to a survey, fewer data are taken because more is known about the test unit, including its airflow rate.

Results include product moisture and any other specified properties that have been measured, as well as equipment condition, especially wall buildup. Weight of product and recovered buildup are taken at points in the system to give a rough material balance and to aid in the analysis of the performance.

A test report normally lists the witnesses and important dates and times, clearly identifies the feed makeup and operating information, states how well the objectives were met, and includes a summary sheet of results. Clients generally appreciate comments on the suitability of the drying method and suggestions for future work.

9.8   Administration

Scheduling.   Effective scheduling optimizes the use of equipment and personnel. As much as possible the order of the tests must consider cleaning and other functions, for example, by placing soluble light-colored products before those more difficult to clean. Schedules can be upset in various ways—feeds not arriving on time, test programs that have to be extended, and other rearrangements and postponements. Formal confirmation letters that commit both the manufacturer and the client help avoid schedule breakdowns.

Maintaining confidentiality.   Every reasonable effort should be made to separate competitive clients, their tests, and their material containers. It helps if feed is not sent too far in advance of test dates. Afterward all containers should be shipped out promptly, and until then kept out of sight. Client names and their products and processes should be kept confidential, and visitors' access to other areas, such as shops and offices, should be controlled.

Regulations and restrictions. In the United States the disposal of all substances is covered by law. Currently many laboratories return to clients all test materials, including wastes and scrubber effluent—sometimes even wash water. Some local governments monitor plant effluents on a regular basis, and many states mandate a *birth-to-grave* responsibility for tracking all materials produced. The generator (original producer) fills out a manifest, a copy of which must be carried by every holder of the material, or any part of it. This includes transporters and disposers, who must be licensed for that class of material.

It is also a U.S. requirement that the generator prepare the Material Safety Data Sheet that lists all the material's pertinent data, including its correct chemical name and physical data; hazard data on toxicity, health, fire, and explosion; reactivity data; spill and leak procedures; and handling and storage precautions. Penalties for infractions of the laws and documents are severe.

Test records. Well-documented test data are essential. They provide a reference and they aid later work, setting initial run conditions in particular, thus shortening test work or making the available time more productive. Useful categories are listed below; some items can be taken from the test questionnaire of Fig. 9.1.

Test number and date

Personnel involved in the testing

Client name, address, telephone number, and test witnesses

Product trade and chemical names

Percent feed moisture (or solids)

Drying medium

Type of solvent

Heat source

Best drying temperatures

Dryer type or name and special features and settings

Comments, including degree of success

Location of samples

A preferred record system lists chemical and other dual names in both the normal and the reverse order, such as calcium chloride and chloride, calcium; corn starch and starch, corn. This arrangement simplifies finding data for exact items and for related materials.

Computer database programs permit sorting by client name, product names, solvent, dryer type, and other categories. Errors and information gaps inhibit use, so the system should be maintained by experienced personnel. The temptation to include installation, sales, or other data should be resisted. Large heterogeneous systems tend to be error-prone, give dubious information, and lapse into disuse.

In small companies data systems are hard to establish successfully—management support may be lacking, databases are relatively small, and employees with long memories have little need for them. But newer employees are less effective without good historical data, and this is made clear by the eventual attrition of people and records.

9.9 Comments

No dryer manufacturer could operate without testing, and it is just as essential to users. While testing benefits from the latest proven methods, there are diminishing returns in greater diversification and more equipment. These may enhance users' opportunities, but they cannot assure the manufacturer's vitality and continued service to industry. The optimum is a balance between the ideal and the practical.

Success depends on a laboratory's ability to get reasonably accurate data quickly, using approximate methods. Failure is a constant threat, sometimes with success almost in sight. Technical problems may dominate, but even in a testing environment the hardest problems are often scheduling conflicts, feed materials not received in time, breaches in confidentiality, and other human errors.

Used Dryers—A New Look

10.1 Introduction

We may not like to think about it, but we reuse water, and we breathe over again the same air that others have used. Yet we resist using castoffs, especially if they are not our own.

The stigma on second-hand things—clothes, cars, furniture—is extended in some degree to the equipment in our plants. Economic necessity, however, is overcoming throwaway habits that might otherwise stick for generations. Many firms are taking a new look at old items of equipment. They have found that their own redundant equipment can be put profitably into some other service, and that used equipment can be bought from others to meet some of the needs once filled only by new purchases.

Recover, restore, and reuse. Reusing equipment is a common practice within a plant or company, but much less common between companies. Nevertheless, many dryers have been resold and reinstalled, some on several different products. This includes large dryers, too big to be moved in one piece. Systems with many components and with drying vessels that have to be cut apart and rewelded are particularly hard to relocate.

In spite of the problems, a number of dryers have been resold three or four times, proving that it is cost-effective. One of these has a 28-ft (8.5-m)-diameter chamber. After having operated for several years at a major chemical plant in Ohio, it was sold to a custom drying firm and moved to New Jersey for test work and contract drying. Later it was sold again, and for a few years was used to dry humate in Florida. At last report it was sold once more, still in good condition.

While the moving parts of a drying system will need repair or replacement, drying vessels themselves seldom wear or corrode if made of the proper alloy. Stainless steels, for example, suit many corrosive jobs. An 18-ft (5.5-m) custom spray dryer operated on a wide range of products for 62 years and had only one modification—lining of the concrete chamber with stainless steel. A 20-ft (6.1-m) dryer at the same plant operated for 34 years and had the same chamber alteration. Even some dryers made of carbon steel have remained in service for many years without much change. One with a ¼-in (6.4-mm)-thick carbon steel chamber has been drying urea formaldehyde resin for more than 45 years.

Changing attitudes. Severe competition has disrupted much of industry and has changed buying attitudes. At many plants there has always been a *boneyard*, a place for storing old equipment, lengths of pipe, and other miscellanea. The benefits are no cost, no approvals, and no waiting for delivery. But it has always been a minor activity. A much more concerted effort is underway, especially among large companies, to recycle old materials, including process equipment. Firms that once had new-only purchasing policies have turned to used equipment as one more aspect of improving their return on investment—but some are still wary. Used-equipment dealers insist that there are no longer any major firms in the U.S. chemical process industry that do not buy some second-hand equipment.

10.2   Coordination of Recycling Efforts

The Investment Recovery Association is a U.S. group that fosters better managing of surplus assets, including equipment. About 200 major processing and other manufacturing companies are members. It promotes recovering, reclaiming, and recycling all usable materials and aims to improve members' profits. Twice-a-year meetings provide a forum where members can learn from one another. They have a wide range of interests that include recovering materials, marketing of equipment, and encouraging the programs among members.

An advanced recycling operation. Typical of the association's members is a major chemical company in the central United States. It practices responsible, yet cost-effective recycling, and has established an Investment Recovery Group at each of its main divisions. Its chemical division has 15 to 20 employees on this work at its headquarters, and about a dozen at other plants. The work is divided into equipment, small items such as pipe and small parts, and metals and other recyclable materials. Yearly turnover is about $10 million, based on replacement value.

For an equipment user to succeed in recycling, cooperation is essential. One of the biggest challenges is to convince young employees, engineers especially, of the benefits of reusing old items on their projects, rather than buying everything new. They have to be educated to an awareness that redundant things can be turned in, reworked if necessary, then put back into use. To encourage this, all exchanging within the company is on a free basis. No price is exacted by one plant for items sent to another, regardless of value.

Company plants in the United States, Canada, and Europe are tied into a computer that tracks all materials. Data from any operation being dismantled are fed to the computer at headquarters where disposal is coordinated. Searches only require typing in the name of any item. That yields complete information on quantity, locations, and specifications. Thus if a plant needs a dryer, tank, or valve, all that are available can be found at once. Commonly used items in stock are listed continually on computer monitors throughout the plants. Operators, in particular, are encouraged to watch for available items that their plant can use.

The reuse of materials within the company is emphasized, because it brings a return several times greater than selling to others. Outside sales account for less than 10 percent of the group's annual turnover. Even so, these sales are actively promoted when inventories are high. If not claimed for internal use in about 6 months, an item is listed in a promotional catalog prepared at headquarters. This is mailed to a list of 3000 to 4000 prospects, including the company's competitors, many of whom are engaged in the same efforts.

Prices are negotiable within a range. Materials not sold to other users in a reasonable time are sold to dealers, which brings only about 25 to 30 percent of the original value. In the case of equipment items, dealers resell for about 50 to 60 percent of current value (normally greater than the original price due to inflation).

Recycling network. User-recyclers cooperate with dealers, contacting one another for needed items. While these users do not buy for stock, they sometimes buy items as a service to regular customers. This helps establish their reputation as good, consistent sources, and avoids losing sales for other items.

Some of the firms, both user-recyclers and dealers, tend to specialize to an extent, and become known for what they have available. For instance, users that make organic chemicals often have good stocks of special alloy items. On the other hand, dealers may have better, more varied stocks of dryers, glass-lined vessels, and reactors. Most large to medium sized dealers handle dryers. Some keep smaller sizes or specific designs in stock; others have larger sizes on consignment.

10.3 Disadvantages and Advantages

Bases for decisions.   Used-equipment decisions have to be considered from many aspects, but usually paramount are safety, suitability, and price. These depend on age, design, condition, complexity, previous use, demand, and other factors. Age extends from recently made to old and run down. It may be a current, or at least acceptable, design, or it may be obsolete for the intended duty. Its condition depends on such factors as corrosion, vibration, metal fatigue, previous duty, and how well it was maintained. Sophistication can range from a tank or other vessel with no moving parts to a complex, highly technical apparatus; and whether a single unit or a system of diverse components, parts may have to be modified, rebuilt, or replaced.

Guarantees.   Even new purchases incur risk, but with used dryers there are added uncertainties. One is the lack of a functional guarantee—operating continuously at a stated production rate, heat load, and power use. For some applications the dryer also has to meet product specifications. Although manufacturers at least verbally imply some acceptance of a functional guarantee, dealers cannot. They have too many dryer types and no facilities for testing. In addition, only a few dealers have the experts needed.

A functional guarantee is less important if the dryer's previous duty was similar to the proposed duty. This is more reassuring than a new-equipment guarantee based on tests on a small unit, especially if the test unit was not a prototype. Of course, capacity and performance are less important when the unit will be used for testing or pilot-plant duty.

In contrast to the guarantee of functional performance, mechanical performance is guaranteed by both dealers and manufacturers. Dealers' customers are invited to witness mechanical checks. Motors, blowers, and other items that fail to operate properly will be replaced. User-recyclers are not able to offer any guarantees, but may take back equipment that proves to be defective. When equipment is returned, the buyer generally must pay all shipping costs, and often receives a credit on future purchases rather than a cash refund.

Leniency by any seller in questionable cases depends on the circumstances and on the buyer-seller relationship. On one occasion some large filter presses were not installed until several months after delivery. They had passed mechanical tests before shipment, but failed in plant operation. The dealer accepted them back even though it was 2 or 3 months beyond the guarantee period.

Documentation.   Information is needed on all but the simplest equipment. This includes operating and maintenance instructions and descriptive information on the main components, possibly with a bill of materials. In addition, drawings may be needed to make repairs or modifications. If not with the item, these are, in some cases, bought from the manufacturer, together with operating and instruction manuals. Dealers' shops can handle most simple repairs, but major alterations have to be made by the manufacturer or some other firm with similar facilities.

Availability.   Much redundant equipment comes to the used-equipment market. Short profit goals and the pace of technology bringing out new products make some operations quickly obsolete. Additional items have become available by the restructuring of the chemical and other industries, items that might have remained in their original service much longer. Even so, it is seldom a buyer's market, and choices are usually limited—not every type and make of dryer is available. Large dealers handle every major type of dryer, but the search for a specific type and size may be fruitless.

Dismantling cost.   The cost of taking down a dryer depends on its size, design, and location. Dealers weigh the value of the system, the potential for selling it, and the take-down cost. Sometimes they will be paid for dismantling, or be offered the unit at no cost for taking it away. If no buyer is found—always a possibility—the item will have to be sold for scrap. Stainless-steel price fluctuations add to the risk. Equipment takedown and reerecting may cost 25 to 50 percent of the dealer's price, but this varies widely with local conditions and with the user's policies—particularly on safety and control instrumentation.

Delivery time advantages.   There are, of course, rewards in buying used equipment, mainly low price and fast delivery. Sometimes delivery is more important, as when a plant has to be running ahead of competition, or when there is only a brief market opportunity. In one such instance a drug company needed a reactor as an addition to an existing process. The wait of 16 weeks for a new unit would have forfeited 1 to 2 million dollars in sales, so a used reactor was bought and installed, easily within the time limit.

Fast delivery also makes it possible to enter a market quickly. If delays are long, startup may extend beyond the ability to predict market conditions. A process cannot be justified if the system takes too long to buy, install, and start up. Some new dryers have delivery times of up to 1½ years, which can put startup at 3 to 4 years for complex plants, in which the dryer is dependent on other process steps. Even allowing for takedown, delivery to the job site can save 6 to 10 months—more for very large designs.

Price and financing. Dealers insist that while cost may not be a big factor in the short run, it always becomes important in the long run. Lead time is primary in some instances, but with used equipment most purchase decisions turn on price. The savings usually balance the various disadvantages. A large-diameter direct dryer may sell for 25 to 50 percent of the price of a new system, although dismantling reduces the savings. Reinstalling is seldom trouble-free, even for a new dryer, and is more difficult for a used one, particularly when modifications are needed. Parts such as controls may have been stripped from the system, and, in addition, one or more major components may have to be changed.

The larger dealers can offer financial assistance in various ways, from loans to lease-then-buy options and various creative financing plans.

Complexity. Used dryers' second or third time around is often for simpler duties. Sometimes they are scrapped after toxic, nuclear, or other hazardous work. But they can become a part of more complex operations. In one case a mirror-image pair of 3500-ft^3/min (1.65-m^3/s) spray dryers first performed a relatively simple open-cycle drying of toner used in copiers. When that job ended, the next duty was for closed-cycle drying of a fine chemical from a solvent.

To fit the system for leakproof operation, first the chamber was accurately rewelded. New fans with shaft seals and better-quality solvent-resistant gaskets were installed throughout the system. For solvent recovery, advanced computer controls and a refrigeration package were added. The second of the pair of dryers was to have been a spare, but was sold when it proved to be unnecessary.

10.4   Used-Equipment Dealers

There are a half-dozen major used-equipment dealers and many smaller such firms in the United States. In addition, many brokers get buyers and sellers together, and a few brokers stock items used frequently in their areas. Some of the small firms specialize in one or two fields. Several, for instance, limit operations to equipment for ceramics processing. The larger firms cover a wide range of fields and stock a variety of materials, but some also specialize in one or two fields. Others carry more sophisticated equipment, partly because they have the in-house expertise. A wide range of dryers and auxiliary components are stored in their yards or at plants on consignment. Dryers comprise about 15 percent of all used process equipment sales.

In various other parts of the world there are some smaller dealers, but most of the activity is carried on by brokers. These operate in South and Central America, the Far East, Australia, Canada, Israel, and some other areas. Many of these sales are made by U.S. dealers and comprise about 20 percent of their total business.

Expert help. Purchasing used equipment is made easier by consultants, who can help with most aspects from search through start-up. In addition, larger dealers have experts on their staffs or on call who give technical assistance. More than in any other way this illustrates dealers' greater sophistication and improved service. One U.S. dealer, for example, has a plant-design consultant who, often in cooperation with other dealers, puts together complete plants. Other experts are on hand to advise on dryers, filter presses, evaporators, centrifuges, and glass-lined vessels. They are able to calculate performances and advise on specific designs and their features and applications.

Cooperative efforts. Dealers have a kind of network, contacting one another to find anything from a small part to whole plants. Even though competitive, they cooperate, especially in handling diverse jobs. During slack times, less cooperation is needed, because more equipment and whole plants are available.

There is also cooperation between dealers and user-recyclers. This simplifies the search for equipment and reduces the need to canvass a number of dealers. Some larger equipment buyers extend this to what is called relationship vendoring, that is, establishing a kind of partnership with a dealer. Both firms get to know each other's methods and requirements, buying and selling from each other. Thus the dealer can give better service by stocking items frequently needed, possibly anticipating needs. And the dealer buys some of the user's surplus.

On major projects, some buyers send flowsheet drawings to dealers for suggestions, for submitting bids, or for reports on the availability of items. Users keep prices competitive by checking other suppliers at times.

Rentals. Renting a dryer is ideal for testing a product that cannot be taken from the plant site or that needs a long-term evaluation. It is also a way to ease into production with a low investment. If the test dryer (or another unit) is purchased, rebates of a portion of the rental fees can often be arranged. Monthly fees for rentals are usually about 10 percent of the selling price with a 3-month minimum.

Some dealers—as well as most manufacturers—rent dryers and other equipment items. Dealers are not equipped to run tests, but may have working arrangements with firms that do. An important part of rental operations is a shop to repair or modify the rental units, and to clean them at each turnaround.

Manufacturer participation. Aside from rentals and an occasional re-sale of a unit taken back for some reason, dryer manufacturers coop-erate little or not at all in used-equipment efforts. Nevertheless, some of their work edges them partially into the used-equipment market. Most of them rebuild their own dryers profitably and supply parts and manuals when their units are resold. They also take on engineering jobs, such as recalculating and redesigning units for new duties.

Dealer and manufacturer cooperation. Reluctance of new-equipment manufacturers to enter to a greater degree into used-equipment oper-ations is thought by at least one large dealer to be misguided. He ar-gues that the number of projects approved is small compared to those dropped, most for reasons of economics. More would succeed, he says, if there were cooperation between new and used suppliers, and all profit factors were considered. Manufacturers would lose sales of sys-tems—often at low profit margins—but they could gain the sale of ret-rofits, spares, and later replacements, which are more profitable. Other possibilities are future engineering and modification work stemming from the saving of otherwise lost jobs.

10.5 Buying a Used Dryer

Getting help. Chapters 1 to 4 described the complexity of drying and the diversity of drying equipment. There are at least a dozen promi-nent types, numerous lesser designs, plus occasional new develop-ments. Each has its own operating characteristics. Thus no individual can know more than a few dryers intimately, which is why help is usually needed to select the right one. Among the first decisions to be made are the type and make of the dryer that is best for the duty, and the components and spare parts needed for the system.

Before a dryer is purchased there is usually an exchange of infor-mation, which differs depending on the supplier—maker, dealer, or re-cycling user. Manufacturers can offer a depth of information, but it is restricted largely to their own dryer types. Although somewhat bi-ased, much of this *free engineering* is useful and not available from other sources. Most manufacturers provide testing services on a fee basis, and the information gained—if paid for—can be used for any purpose.

Aid from dealers is broader and less biased, but is limited, because they stock diverse supplies as well as equipment, and they do not have much depth of engineering talent. Dryers are only one of many kinds of equipment, and these vary in type and in makes within each type. Even the experts on the staffs of larger dealers do not have full knowl-edge of all kinds of dryers. Recycling users, in contrast, offer even less information, although they may be more likely to have drawings available. They have less incentive to promote sales by providing information, because their main effort is to recycle within their own company. When selling outside, they must protect proprietary company information.

The search. If not well informed about dryers, a purchaser should consult an expert, especially if product drying tests will have to be run. Help is also available from manufacturers and dealers. When the need is for a complete system, a specialist on a dealer's staff can help to put it all together, using the informal network of dealers and recycling users. A project engineer may be needed for replacing unusable or missing components. Some such items can be found by the dealer, but more have to be purchased new.

Locating a single item is much simpler than assembling a complete system, for which an extensive search may be needed. Parts of systems are often cannibalized for sale or use elsewhere. Controls and motors are the most likely to be stripped.

Outfitting a used drying vessel to suit a new application sometimes requires a number of different components. If most of these have to be bought new, the system is used in name only. A set of two 10-ft (3-m)-diameter spray dryers, for example, when resold were furnished with mostly new equipment—centrifugal atomizer drives, air filters, heaters, bag filter collectors, blowers, structural steel, and computerized controls. It was essentially new, but installation was faster and at a much lower price than a totally new system.

If a large drying system is to be bought from a dealer, it should be a firm with large resources and access to stocks of other dealers and user-recyclers. Such a dealer can put together most of the components, which will be on hand or can be located. Even so, it is advisable to hire a dryer expert if the dealer does not have one available.

Mechanical details. There are many mechanical factors to be checked on dryers, which often must operate under hazardous conditions. Safety is primary. What are the chances of fire or explosion, for example, and what measures should be taken to avoid or minimize them? Can the effects of wear or damage be corrected, and by whom? Will leaks cause problems? Most dryers, if not specially built to be airtight, leak to some extent, even when new. A particular difficulty with used dryers is contending with parts that are corroded or out of shape.

Drawings, bills of material, and manuals—often not included with a used dryer—may in some cases be bought from the manufacturer. If not, they may have to be redrawn and rewritten. Otherwise it will be necessary to do without them, which makes repairing, installing, startup, and operator training more difficult.

Large idle dryers are often left standing at the plant site, and are moved only after having been sold. Dismantling can be a major undertaking, especially so for large dryers fitted indoors into tight places. A few firms specialize in takedown work, but erecting and installing is normally not much different than for a new system. Problems of responsibility may arise if a unit is not rewelded by the same firm that cut it apart for shipping.

Inspection.   The main purposes of inspection are to assure that nothing is missing, that the equipment is what it is represented to be, and that it will operate properly. Drying vessels in particular should be checked internally for corrosion, wear, or points where leaks could occur. Safety considerations are a top priority. Depending on circumstances the inspector may have to judge the value of the equipment, the need for repairs, and the usefulness of manuals, drawings, and other documentation.

Motor and fan rotations, hydrostatic and vacuum tests where required, and general mechanical checks are made by the seller. Electrical characteristics should be noted—one U.S. standard is 60 Hz, while Europe and many other locations use 50 Hz. Tests for stainless steels can be made roughly with magnets, or more thoroughly with chemical testing kits.

Functional operation.   If operating conditions are not known, before committing to a purchase, the buyer should have a test run on the specific product. An engineer or technician experienced in dryer operations should witness the tests and help to evaluate results. Tests to assure product quality are generally run on small dryers. These can be made on rented test dryers or at a test laboratory, such as at a manufacturer or custom drying firm. The scaleup of results to the desired size is different for each dryer type. These calculations confirm the production rate and heat and power loads, and should be made even if confirming tests are not run.

Procurement.   As noted previously, used equipment generally sells for one-half the current new price, somewhat less for dryers with large chambers. Much depends on the unit's condition and on supply and demand. As experienced buyers know, and others must surely suspect, dealer prices are always negotiable, within reason. But that is also true for new dryers, at least large ones. Some buyers benefit from getting help with price negotiations, guarantees, and creative financing plans.

It is important to establish legal ownership. This is occasionally a problem, because the title to some equipment or entire plants may be in dispute or actually in litigation. In other cases title to ownership is not firmly established or is passing from one owner to another.

Table 10.1 summarizes the points made in this section. For the best chance of success it is good practice to get experienced help to advise, inspect, investigate, evaluate, witness testing, and perform various other functions. Some of the operations required for completing a purchase successfully are detailed in other chapters.

TABLE 10.1   Items to Consider when Purchasing a Used Dryer

Educational	Dryer types and capabilities, system components
	Spares required
	Used-market conditions
	How to get engineering and other help
Mechanical	Inspection of all system parts
	Mechanical checks: all moving parts
	Mechanical safety aspects
	Drawings, bills of materials, and owner's manuals
	Damage: effects of wear, corrosion, vibration, poor operating practices
	Repairs, replacements, or modifications
	Dryer takedown and relocation
Functional	Safety: fire, explosion, toxicity, pollution
	Testing: production rate, fuel and power use
	Design calculations: production rate, fuel and power use
	Operating instruction manuals
	Startup
	Training operators
Procurement	Suppliers: dealers, users, manufacturers
	Dryer search
	Search for components
	Investigate ownership
	Value and price
	Purchase negotiations, financial arrangements
	Guarantees

11

Commissioning

11.1 Introduction

Commissioning a complex dryer is more than just a start-up. Of the various phases of producing a complex system, it is probably the most challenging and requires the broadest range of skills. Because it is the focal point for any errors made in earlier stages—testing, design, and manufacturing—it is nearly always the supplier's responsibility. Who better to find and rectify whatever needs correcting?

Many dryer "manufacturers" are actually engineering firms that design the drying system and contract out the fabrication. For most dryer types the vessel and connecting ducts are fabricated of light-gauge sheet metal. Standard components are bought to complete the system, and all items are shipped to the job site. Responsibility for erection and installation, on the other hand, is often assumed by the purchaser.

The original start-up of a simpler dryer is quoted separately and may be handled with no help from the supplier. The purchaser is likely to contract some of the installation and start-up out to local firms. Much depends on the type and size of dryer and the know-how of the purchaser. Small to medium sized indirect dryers in particular are relatively easy to install and start up.

A small, simple dryer might take one experienced engineer or technician a day or so to start up. Most medium-sized dryers require 1 or 2 weeks. But large systems need extra help, especially those on more than one level or running more than one shift. Thus a full commissioning can last 4 to 6 weeks, even longer when there are unexpected delays, or if the system includes steps other than drying, such as feed pretreatment or product conditioning. Furthermore, when the dryer is part of a large plant, complete commissioning often takes more than a year. Final drying is generally the last of several processes, and its completion is slowed by any problems in prior operations.

Individuals who commission dryers may be engineers or knowledgeable technicians, but are referred to here as technicians or the supplier's crew. This is to distinguish them from the user's start-up engineers and technicians.

11.2 Erection and Installation

There are two steps before commissioning—erection (foundations, support, and access) and installation (mainly wiring and piping). Most often, contracts for both steps are let by the purchaser to outside crews; thus the supplier has little or no control, and that obscures responsibility for any problems. To help avoid misunderstandings, the technician should contact the plant at regular intervals and monitor the progress of major phases.

There are sometimes difficulties fitting process equipment into its supporting structural steel, such as a beam blocking the path of a duct. Such interferences usually result from drafting errors—trying to show three dimensions on a two-dimensional drawing. They are less apt to occur if the manufacturer has a scale model built, or makes the structural drawings in house and thus can review them more often. It is common practice, however, to have the structural elements drawn by an outside specialist.

11.3 Technicians' Preparation

Taking a complex new system into full production is a multistep effort, and to be done well it needs close cooperation between buyer (user, customer) and supplier (manufacturer, vendor). The connecting links between the two are their respective start-up engineers or technicians, and if they work as a team, both sides benefit. The various steps they perform in commissioning a typical drying system are listed in Table 11.1.

Before going to the plant, the supplier's crew should list all the operating conditions, major vendors, and other contacts, and assemble any relevant items not already at the site, including drawings, equipment list, and operating instructions. The manufacturer's engineer, who prepares these instruction manuals, should cover all steps in the system's operation, including start-up, shutdown, and all precautions. They should be detailed enough to instruct new operators, making allowances for high turnover rates and job changes within organizations for operators and supervisors.

To be able to teach plant personnel, the supplier's crew must themselves understand all the equipment in the system. Of special importance are burners, controls, and items with recent design changes. It

TABLE 11.1   Commissioning Outline

Precommissioning duties	Review design
	Record pertinent operating and design criteria
	Discuss plant readiness; work from a list
Initial start-up duties	Conduct initial discussions
	Inspect system; start checklist
	Make in-plant contacts with contractors
	Check electrical-mechanical items:
	Motor rotations
	Feed drive
	Safety controls, including burner
	Feed conditioning or preparation equipment
	Total system review, including panel board and instrumentation
	Begin operation:
	Set start-up procedure
	Set shutdown procedure
	Train personnel
	Take performance data
	Amend and revise operating instructions
	Turn operation over to supervisors and operators with completed checklist
	Observe operation
	Instruct operating maintenance personnel

may be necessary to get instructions on some items from colleagues or vendors. This know-how is needed, not just for operating, but also for making settings or adjustments—sometimes even for making modifications.

Both crews should know each other's applicable company policies and be informed on the plant's labor situation and progress of the installation. For instance, it helps to know the status of the plumbing and wiring, and whether blower and motor rotations have been checked.

Keeping well-informed will also reduce the likelihood of the technician being summoned to the plant before it is ready, sometimes only to meet a preset schedule or to help with installation problems. The temptation is strong, when either advice or assistance is needed, to call in the supplier.

Inadequate contact marred the commissioning of a large high-temperature spray dryer in Jamaica, West Indies, somewhat remote for regular telephone calls. It had been agreed that, with no help from the manufacturer, plant personnel would be responsible for erecting, installing, and starting up the system. Unfortunately, during the installation they assumed that the sliding fit of the hot air duct's expansion joint must be a mistake, so they welded it closed. As a result, after a few temperature cycles from start-ups and shutdowns, the dryer's roof buckled.

11.4 Initial Duties

When they get to the site, the supplier's crew should be briefed on any items not previously covered, including work rules. Following that, a plant walk-through will give them a more complete picture of the system, which they have known only as drawings. Then the user's engineers can fill them in on any other operations that affect the drying system. Throughout, they all should look for any oversights or errors that need correcting.

Mechanical and electrical checks. Next, mechanical and electrical items ought to be checked, giving special attention to blower speeds, motor rotations, safety controls, and instruments. It may be necessary to make adjustments to the burner, if there is one. To assist on major or complex items, vendors sometimes offer to send in their own technicians.

Bag collectors. On pulse-type bag filter collectors it is best to wait before turning on purge timers until a mat of particles has accumulated on the fabric. At this point pressure drop across the collector is about 5 in WG (1.2 kPa). Then, setting the initial pulse interval at the shortest time on and longest time off will least upset the drying airflow and use the least compressed air. As the system reaches continuous operation, adjustments can be made to the pulse-timer interval, if necessary, to maintain a constant pressure drop through the fabric.

Rotating equipment. Examining impellers of blowers and rotary air locks for tramp metal will help prevent sparks that could cause a dust explosion. They should also be checked for both speed and direction of rotation. If the sheaves on a belt-driven blower are reversed, the impeller will run faster (instead of slower) than the drive motor and may cause the impeller to shatter.

11.5 Actual Start-Up

Safety. The first consideration must be safety and proper regard for the plant's policies and procedures. Evading plant rules can boomerang and hurt good relations, safety, and even ultimate costs. In some plants a simple job may require a millwright, plumber, or electrician, or all three, sometimes with crew supervisor and helpers, to do what a start-up technician might be tempted to handle alone. Nevertheless, the smallest incident can generate a grievance report or otherwise erode the team effort and make it harder than ever to get things done.

Modifications. Modifying standard purchased equipment items may void warranties and actually impair performance. Unauthorized field modifications may have to be made if there is great urgency, or where access and deliveries are poor, as in some third-world countries. But before taking chances on alterations, it is best to get the vendor's advice if at all possible.

Shortcuts. Taking shortcuts is risky, although at times considered necessary by experienced technicians. To avoid holdups caused by control systems, for example, they may temporarily defeat one or more of the safety devices. It is important that a system never be left in a bypassed mode, however, unless everyone involved is aware of both the risks and the corrective steps that may be necessary.

Before turning on any equipment, all the consequences must be understood. For instance, if the system has a bag filter collector, watching the air temperature and allowing it to rise gradually will help avoid damaging the fabric. Lighting a gas or oil burner needs particular care. Only properly trained personnel should make the first lighting of even a laboratory-sized unit.

Turndown. Direct-fired systems that have to retain full air velocity (to assure conveying the solids) need a burner with a turndown ratio at least able to match any reduced-capacity requirement. For this situation the preferred control method is to have the feed rate regulate the inlet air temperature (otherwise control during reduced capacity has to be manual). On start-up this occurs when feed cannot be supplied at the design rate, and also when proving a turndown guarantee.

11.6 Operating

Getting up to temperature. Details of starting the dryer are given in the operating instructions, and safety switches will prevent the most common errors, such as feeding the dryer without first turning on the blower and heater. But one aspect of operating that requires developing a special skill is stabilizing the air temperatures and feed rate. For a cold start with a fired heater, typically the dryer inlet air temperature is gradually increased for a period of about 30 min, giving each part of the system time to warm up properly. (A heater with refractory lining, however, may take several hours to reach equilibrium.)

For most direct dryer types the outlet air temperature is held at about 20°F (11°C) above the design, then gradually lowered as the inlet temperature rises to its design point. Because of this initial step, some product may have to be reworked or discarded.

The procedure for spray dryers is the same as described, except that, initially, just water is fed. The outlet air temperature is kept above the design by about 20°F (11°C), possibly more for feeds with high solids. This helps to avoid either splattering the chamber or plugging the line to the atomizing wheel or the nozzle. Feed is slowly substituted for water when the dryer inlet and outlet temperatures have properly leveled off. Then the outlet temperature is gradually dropped to the design. (Conversely, the dryer is shut down by easing from feed to water, gradually bringing up the outlet temperature.)

11.7 Problems

A dryer's first start-up is sometimes held back by problems with fabricated items, but after the job has begun, most delays result from changes required by the initial electrical check. Computerized control systems, common in large plants, also cause some difficulties in control sequencing. Major ailments of system performance, in the areas of product characteristics, production capacity, and condition of the equipment (including leaks) are covered in Chap. 12.

When drying is the last of several steps, the commissioning time depends to a large extent on success with the previous steps. Two examples are feed either held up or not at the specified solids content or rate, and operators or craft workers not available because they are assigned to something more urgent.

Feed differences. The system design is based on tests run on a specific feed. But feed from the plant may be different, forcing changes in operating conditions, or possibly changes in the equipment. To illustrate, the moisture content of a wet solid may be so high as to require backmixing at a higher rate than the design. Liquid feed may have to be diluted, otherwise altered, or heated to make it flow, spray, or dry as expected.

Getting on stream. A major concern of plant managers is to meet the target date for getting the plant operating—delays hurt earnings. Under most circumstances it is best to defer trying for full production rate, product properties, and clean equipment conditions until later. Instead, get the plant on stream as soon as possible. It is easier to concentrate on fine-tuning when pressures have lessened. Raising productivity and lowering the cost of operating are covered in Chap. 8.

11.8 Taking Data

After the system is running, the technicians should record the operating data, most importantly the data that relate to guarantees. While taking data is a chore, there are seldom too much and often not enough. They will also be useful later to compare against the earlier test results and perhaps help in future work for this or other applications. But certain process conditions, if developed by the user, are considered confidential, including some of the data.

Table 7.1 lists the data taken in surveys, and the same items apply to start-ups. Some are difficult or impractical to get; the minimum are those needed to confirm guarantees. Data are also needed for some dryer types to allow the calculation of loading and exposure time. It may be necessary to take dust samples at the dryer exhaust stack and measure the airflow. These are needed for calculating dust loadings and losses for meeting federal or local regulations, or to confirm collector performance.

11.9 Guarantees and Other Agreements

After the plant is running properly, any guarantees or other representations have to be proven. Most of these concern capacity, run duration, turndowns, equipment condition (such as product buildup), product characteristics, and operator and maintenance training. Filing for a license or for approvals to operate the system in some localities is another requirement, but is usually the owner's responsibility.

The guarantees to be proven during commissioning should be based on the material that was tested. Some allowances have to be made if the tests were run on a simulant. The actual feed may not have been available when tests were run. When the actual material was not tested because of its hazards, start-up personnel should not be exposed to it. They are even less likely than the test personnel to be qualified to deal with such dangers, especially considering the unregulated, and sometimes hectic, nature of start-up operations and the larger quantities of material handled.

Production-evaporation relations. A direct dryer's capacity is a function of the evaporation rate. The product rate, however, is of most interest to plant management. So the technician must understand the relation between product rate, evaporation rate, percent feed moisture, and product moisture. This is explained in Sec. 7.6, using Eqs. (7.1)–(7.3).

To illustrate, a solid feed with 30 percent moisture yielding product with 2 percent moisture would have a product-to-evaporation ratio of

2.5. If the feed moisture rises to 32 percent, the ratio drops to 2.267, causing a drop in the product rate of 9.3 percent.

Improvement strategies.   If the system fails to attain the design evaporation rate, certain optimization techniques may help, such as improving the dryer's inlet and outlet air temperatures and heating the feed. (Refer to Chap. 8 for details on this and other strategies.) But because these efforts are time-consuming and diverting, they are best put off until the essential parts of commissioning are finished.

Product specifications.   There may be verbal or written agreements referring to product characteristics such as moisture content, bulk density, and particle size and range. Some specifications, however, are hard to meet if based on small-scale tests, because product dried on a large unit goes through a longer and somewhat different action.

In addition, product properties are affected (more or less, depending on the type of dryer) by the feed properties. If the feed moisture is too high, for example, the product may be unacceptable, with regard to bulk density and particle size, in particular. Discrepancies in product specifications will confirm the benefit to both supplier and purchaser of keeping good written records and retaining samples of feed and product from test runs. Any changes in feed composition can make a great difference in the operation of a dryer, sometimes making it ineffective.

Specifications vary widely from one product to another. For a dairy plant, standards are set for taste, moisture content, bulk density, and absence of brown specks. Dried coffee would have similar standards, including a taste test and examination for black specks. Other products might have to be tested for various properties: a fluid-bed cracking catalyst for particle density and size, as well as bulk density; a pigment for nonagglomeration plus a smear test for grit and coating ability.

Extended run.   Even if not a requirement, it is very desirable to make a run of a few hours or more. In this way continuous near-normal operation can be confirmed, revealing, among other things, any material buildup. The technician need not be present throughout, but should check the system when the run ends.

For bed-type dryers it is usually necessary at some point to measure the loading factor and exposure time. After shutting down, the volume of material in the vessel is measured and compared to the total volume of the cylinder and the volumetric flow rate. Another method of measuring the exposure time in a low-temperature dryer is to use a colored particle as a tracer. A high-temperature dryer can use a low-level radiation pellet. But these tracers must have size and densities nearly identical to those of the average particle in the dryer. The exposure time distribution is a bellshaped curve, which can be modified using dams and other devices.

Other demands. If it is part of the contract, the system turndown will have to be proven. In addition, there are other, more subjective, considerations—air pollution, frequency of cleanouts, noise, vibrations, and satisfying OSHA and EPA regulations. For some situations other suppliers or consultants have to be called in for data gathering, testing, modifications, or advice.

11.10   Operator Training

Operating instructions. Operators have to be taught a range of skills, from setting control systems to the proper recording of data. The operating instructions are helpful at this time, and will continue to be valuable as an official reference. When the plant is running properly, it is up to the technician to revise the instructions to reflect any changes made to the original design. Afterward, even though they will lie idle most of the time, the plant engineers should keep them updated.

Benefits of training. Just as with taking data, there is a temptation to shortcut the training of operators. A thorough training program has obvious benefits for the user—more economical operation, better product, and fewer expensive mistakes. The supplier benefits as well, although in less tangible ways, because of fewer time-consuming calls for advice and less free service (which is often extended for months after commissioning ends).

Checklists. More likely to be used for operations on a regular basis than operating instructions are checklists, or summaries of operating steps. They are especially valuable for complex or hazardous operations. One can be drawn up for the overall system, others for individual equipment items as needed.

Work habits. If supervision is lax, the best instructions and checklists cannot prevent poor work patterns from developing, and with multiple shifts they can get worse. In time bad habits stiffen and resist correcting. Two general conditions to watch for are taking shortcuts and defeating safety devices. Another is operating too conservatively, such as running at less than ideal feed moisture or air temperatures. These may help to assure clean surfaces in the dryer, but they lower production rates and raise energy use.

Some specific bad practices include erratic setting of valves and louvres, poor maintenance of control instruments, inconsistent setting of temperature controls, and failure to take readings (dials perhaps not well placed). These are often signs of present or future trouble, or at least reduced efficiency, such as lower production rate and higher energy use.

Retraining. Because initial eagerness and cooperation fade, and because job changes are frequent, retraining by the supplier is recommended about every 2 years, more often when indicated. At these visits the technician can conduct seminars for each shift and review the system, the procedures being used, and the operating instructions. It would be unusual if nothing needed correcting or updating, even in a well-supervised plant.

Continued monitoring. After commissioning, purchasers will get a better return on their investment if, on a regular basis, they continue to monitor the feed condition and the operation of major equipment items. Also, system performance should be checked by periodic calculation of the evaporation rate. To avoid the decline in efficiency that occurs in many plants, the important phases of the operation have to be analyzed from measurements of moisture content, temperature, pressure, and flow rate. It helps to have the needed instruments in place, which, in fact, would also expedite start-up.

11.11 Conflict

Commissioning personnel face conflicting demands from the two principals. The supplier wants the fastest possible commissioning, because final payment is customarily withheld until the job is accepted. The purchaser also wants a timely start-up, but in addition has a right to demand good documentation, well-trained operators, and a good running system, none of which can be rushed. Commissioning is even more difficult on projects managed by engineering construction companies, because their objectives are different from those of either supplier or user.

12

Troubleshooting

12.1 Introduction

Troubleshooting is complex because of its interrelated activities, not easily arranged rationally into problems and methods of solution. In some ways it is similar to commissioning, which generally is on a new system, while troubleshooting is on an old one. Commissioning covers the entire process. A troubleshooting job usually concerns one area of difficulty, but it may take a full survey and analysis to find and correct a problem. So while there are differences, troubleshooting is often a modified commissioning—a kind of first aid from the factory. It needs at least a portion of a survey and some of the techniques described in other chapters.

When a problem arises, it is sometimes necessary to use a consultant or third-party technicians to solve it. They may have to establish responsibility when, for instance, costs have to be assigned. In addition to determining what has gone wrong and where, the technician will usually have to find a solution and make recommendations. He or she may have to interview personnel at the site and elsewhere to gain enough information on which to find solutions.

Some of the more common reasons why troubleshooting is needed follow.

1. Mechanical breakdown

2. Poor performance of critical equipment items

3. Product buildup or condensation, or both, on equipment walls

4. Adverse change in product characteristics, such as nonuniformity

5. Change of product to be dried

6. Loss of powder to the atmosphere

7. Need to increase production rate

In many cases taking data, as in a survey, is the central part of troubleshooting. When there are disputes to settle, a formal survey may be required with measurements structured according to a mutually accepted standard. This could be a modification of the "Spray Dryer Testing Procedure" (AIChE, 1988), altered to suit the specific type of dryer.

12.2   Planning and Executing

The first steps in troubleshooting are much the same as for commissioning—become familiar with the operation before visiting the plant. Past history is often a good guide to what needs investigating. Test records for this and similar products should be studied. Instruction manuals, equipment specifications, drawings, inspection reports, and other documentation relating to the system should be reviewed. Finding problems is easier if the safety factors added to the various equipment items are known. It may be unreasonable to expect that all calculations be rechecked, but computer input and output data should be examined. Wrong input is a major source of errors.

At the plant, the technician or engineer should become acquainted with the plant's safety and labor practices. A tour of the drying system and related parts of the plant will provide background information and the chance to look for potential trouble spots. Discussions with supervisors and, if practical, with operators may yield the most useful information on the extent of the problem and its history.

It helps to know beforehand whether the problem began suddenly or worsened gradually, or whether it has always existed. If sudden, the change might have resulted from instrumentation breakdown, product blockage, or maintenance on the system. It is also possible that something unforeseen has occurred in the system. Some foreign material may have been left in a duct or item of equipment. Problems that develop over time, on the other hand, could result from product buildup, corrosion, or other deterioration. If the problem has existed for a long time, it may have been caused by a fault in either manufacture or initial installation.

If indicated by the kind of problem, operating logs and charts should be reviewed, as should maintenance records. Have operating procedures changed, and do they deviate from the operating manuals? Have maintenance procedures or schedules changed? Is the dryer affected by other plant operations, such as reactions, filtration, or other steps that prepare the feed?

A single item of equipment may be the initial focus. But the whole system may have to be studied if the cause is remote from the evidence. Such cases require an overall review of plant operations, just as in a formal commissioning, with mechanical and electrical checks. This includes fan and motor rotations and safety controls. Each type of dryer requires a different approach, but for each a checklist should be followed.

For all but the simplest and most obvious problems, a systematic approach is needed. A logical plan is to identify the trouble, observe and measure relevant operations, analyze the data and other evidence, and make or recommend corrections. There is no detailed procedure, however, that will solve even a portion of the possible problems because of the complexities.

For some specific steps permissions are required, such as drilling holes to insert measuring instruments or making minor modifications. It may be necessary to override a safety to save time or to bypass an item that is malfunctioning or giving a wrong signal. But safety of all personnel has to be the primary concern. Dangerous equipment should only be operated by those who are properly trained and know the consequences of shortcuts. As an example, when starting up fired heaters—even laboratory units—shortening purge times and bypassing safety devices creates risks that should be taken only by experienced technicians and operators.

12.3 General Dryer Problems

The many troubles that beset drying systems are not readily classified, and only some of the most common are covered here. They have been divided into four classes as a convenience; there is some overlap between them. They concern equipment performance, equipment fouling, low production rate, and product quality. Finding solutions usually means taking measurements and making calculations. It requires a knowledge of most of the practical aspects of drying in the specific type of dryer. Some information on these was given in previous chapters.

12.3.1 Equipment performance

Not every problem originates within the plant. The engineer starting up a large spray dryer just outside a small town was able to get it running well, but for no apparent reason the system shut down every morning about 6:00 A.M. Each time the high-speed atomizer drive cut off, it shut down the other equipment in sequence with it. The engineer thought the problem must be electrical, but the third day it happened, he discovered that the pressure switch on the drive's cooling water was the cause. Every morning the water pressure dropped when the townspeople got up to shower, shave, or whatever. To avoid being shut down by outside causes and flow fluctuations, many cooling systems are set up as closed loops. They also conserve water.

An occasional problem with equipment is performance below expectations. Collector efficiency, for instance, may be low. Equipment may also break down or fail to operate as a result of faulty design, manufacture, installation, or use. In such cases the problem may occur independently of, or when operating with, other equipment. An example of improper installation is a blower turning backward; it may still operate at 50 to 70 percent of full capacity. Such an error may go undetected for a long time at a plant operating at less than full capacity.

Safety controls are normally interlinked so that if one item is shut down, it automatically shuts down any others that would affect the operation. In a spray dryer, for example, if power to the atomizer drive fails, the feed pump and burner are shut down. This keeps wet feed from splattering the chamber, and it prevents the drying gases, no longer being cooled by evaporation, from damaging the collector and the blower.

If control instruments are not maintained properly, readings become increasingly unreliable. Temperature controls get out of calibration and pressure control lines plug, until finally the affected readings are meaningless. Production may continue in this way for a long time, but at a decreased production rate and possibly lower product quality.

Many equipment items are designed in size increments, and of the two sizes straddling the requirement, the larger is selected. Then it is further oversized if safety factors are added. An item may get safety factors from up to three sources—purchaser, system designer, and equipment manufacturer. While an overall safety factor of 10 percent or more is no longer a tradition for the design of drying systems, safety margins are still routinely added to avoid bottlenecks—as of course they should be, within reason—to heaters, blowers, feeder drives, and other critical items.

Not all oversized equipment performs as well or better than the design size. As an example, cyclone efficiency drops if the diameter is increased so that the air velocity is below the design range. Conveying ducts, on the other hand, have to be sized to give an air velocity high enough to carry all the powder, but not so high as to cause abrasion or an excessive pressure drop. Too small is never right, but too large is not always beneficial.

A recurring equipment problem is the cracking in freezing weather of water-cooled jackets of cast iron. Unless such units have antifreeze protection, they should be drained when taken out of service. Precautions such as this ought to appear in the operating instructions and be properly emphasized.

12.3.2 Equipment fouling

For equipment to run continuously with the solids flowing smoothly from the feed point to the discharge is an ideal. But fouling can result from sticky material, from nonuniform distribution of wet feed, or from unevenly heated surfaces. It may be that dry material should be added to wet—rather than the reverse—but this is not possible, as when wet feed is added to partly dry material inside a dryer. In bed dryers balling, or the formation of lumps, sometimes cannot be avoided. The solids have to go through that phase, during which time some fouling may occur. Buildup can be minimized if a dryer has scraping devices, preferably adjustable.

Solids buildup. Dry particles drop out of an airstream if the velocity is too low or uneven, a frequent problem in air-conveying ducts. Solids occasionally drop out in elbows that convey powder, such as at the bottom of spray dryers. If the product does not degrade, the buildup may be self-correcting, as it increases the velocity until it is high enough to convey.

To keep the duct clean, a vertical baffle can be installed with an area equal to, or slightly greater than the cross section of the blockage. Placed in the duct above the place that is blocked, it will allow the solids to be swept along the bottom. Alternatively, a temporary slide baffle can be installed and the right depth found by trial and error. Another remedy is to keep the solids entrained with an air lance or an air spoiler that creates an eddy current.

Fine material sometimes touches hot surfaces and fuses in place. To avoid this, a common design includes a jacket over the hot area to allow cooling by air or water. This keeps the metal cool enough so that particles do not stick permanently. But it may be difficult to add such an item in the field.

Sometimes wet or tacky powder builds up in ducts conveying powder. Dried solids can regain moisture where condensation has occurred on cool surfaces or in filter bags. This can result if the air is too close to saturation or, in the case of an outside surface, from inadequate insulation.

The same situation can occur in short-exposure-time dryers that give more time to the product between dryer outlet and collector. The problem occurs either in long ducts or in collectors, such as bag filters, that hold up the powder for a few seconds until it is shaken loose.

Product at the dryer outlet is still somewhat wet. Some sticks there, and the rest reaches dryness only after the added exposure in the ducts or collector. Setting the dryer outlet temperature higher may correct the problem.

A large buildup can partially close off passages and increase pressure drops and power costs. This kind of problem is always possible in a bag filter collector during process upsets. Some accumulation of powder on the fabric is normal, so added buildup may not be noticed for a time, and the dryer capacity drops.

Very difficult problems—usually beyond control, or even the ability to measure—are a dryer's airflow patterns. The main flows of air create turbulence, and vessel design and the drying action affect flow directions and velocities in ways not fully understood. Obstructions create dead spaces, where small eddy currents can carry fines that adhere to hot, or even to cool, surfaces. To know the cause of such problems may not provide a solution, but could prevent an expensive, but futile revamp.

Condensation. Condensation can occur at an air moisture content close to saturation, which is the usual condition at the outlet, and it extends to the collector, blower, and stack. In any of these places condensation can occur where metal tends to be cooler, such as at flanges and where insulation is difficult to keep in place. It can worsen a buildup problem and cause serious fouling of equipment surfaces and possible corrosion. Condensation and corrosion can also occur during long shutdowns. To prevent it, some plants, after shutting down, continue a small flow of warm, dry air. In particular, dryers with carbon-steel surfaces need this care throughout shutdowns.

Systems that cool powder-conveying air to a low temperature are vulnerable to wetting of the dried material. As the air passes over the cold tubes of the cooler, it becomes saturated and water condenses out. To bring the air safely above the saturation point, it is reheated a few degrees. Before reheating, the moisture must be removed, but this may be difficult if the unit is under partial vacuum. When condensed water cannot drain out, a trap is needed to remove it forcibly.

12.3.3 Low production rate

Capacity problems are probably the most common; only a few are covered here. Two of these are low collection efficiencies and low solids content in the feed. Others include leaks, very conservative operating practices, and instruments that have gone out of calibration. A trouble may have only one obvious complaint but several possible causes, which may be simple, or they may be complex with multiple effects.

Collector efficiency, cyclones. Low collection efficiency means that powder is being lost to the atmosphere or to another collector. If carried into a secondary collector, losses are hidden, and some—as from a wet scrubber—cannot be returned to the product, and thus the recovered production rate falls by that amount.

If the air velocity is not held within a narrow range, the cyclone efficiency drops. This can occur, for instance, if the airflow rate is lowered, as when decreasing the production rate. (That would be avoided by lowering the dryer's inlet air temperature instead.) To increase the production rate, on the other hand, either the airflow or the inlet temperature has to be increased. If the airflow is increased past the cyclone's design range, the efficiency drops.

Lower efficiency also results from overloading a cyclone with solids in the airstream, as when the capacity is increased. Severe overloading is possible in systems that air-convey the powder to a relatively small cyclone. A two-stage drying system is similar—cyclones for both dryers must have airflows in their proper ranges. Trying to correct the second dryer's cyclone efficiency by cutting back on the primary's airflow may reduce the efficiency of the primary's cyclone. Thus the pressure drop across cyclones has to be monitored on a regular basis, especially when changes are made in either airflows or product rates.

Typical design pressure drop for high-velocity cyclones in drying systems is 4 to 8 in WG (1 to 2 kPa). If the pressure drop falls substantially below its design, the collection efficiency is lowered. The relation with air velocity, thus airflow rate, is given by the equation

$$P = KDV^x \tag{12.1}$$

where K is a constant determined by the units chosen and the type of cyclone. P is pressure drop, D is air density, and V is air velocity. The exponent x is usually reported as 2.0, but it is 1.75 on some high pressure drop designs. For a specific cyclone, if one condition is known, others can be evaluated.

Cyclones have several advantages over other collectors, and with no moving parts should last indefinitely. But in the real world, operators hammer on cyclone tips to make them discharge better. This damages the tip, and in extreme cases will reduce both the cyclone's efficiency and the accuracy of its pressure drop data. It may also create or worsen leakage, which hurts efficiency. When cyclone tips are damaged, they should be replaced with tips that are reinforced to withstand hammering. Automatic, mechanical vibrators mounted on cyclones can correct slight blockages, but major stoppages will be packed in tighter, so the benefits are mixed.

Collector efficiency, bag filters. For air-suspension-type drying—flash, fluid bed, spray, and some others—the most practical bag filter collector is the jet pulse type. It causes the least fluctuation of airflow, and so lets these dryers perform more efficiently. Collectors that periodically shut down one of several compartments for cleaning are unable to hold as steady a pressure drop. But these collectors have other advantages, and are in common use on other dryer types as well.

Bag collector efficiency is often considered to be 100 percent, except for very fine particles. This is true unless a bag breaks or comes loose from its clamp, or if there is a mechanical failure in the unit's tube sheet. Proper bag selection and installation and regular inspection will help to minimize leakage and to maintain continuous operation with a steady pressure drop.

Some materials hanging up in bag filters can cause fires or deterioration of the fabric. If suitable replacement bag material cannot be found, another collector type or other arrangement may have to be substituted. In most cases this would be a multiple cyclone or a cyclone–wet scrubber combination.

Conservative operation. Another reason for low production rates is operating too conservatively, that is, at less than optimal feed solids or drying temperatures. These safe conditions may keep the drying vessel and ducts clean, but they reduce the drying efficiency. Nevertheless, they may be justified in some hazardous applications to maintain a safer operation.

Calculation errors and omissions. If the production rate is lower than it should be, there could have been an error in the process calculations. Some errors result from the use of computers, which has taken away the visual checks of manual calculations. Reviewing computer input data is an important step to prevent errors and to save time finding them later.

There are several things that may have been overlooked. A few possibilities include plant at a high elevation, heat of crystallization omitted, in-leakage of air with the feed or at some other point, and various other items that should have been anticipated. To show the importance of elevation, for example, if calculations were made for sea level, but the elevation is 5000 ft (1524 m), a dryer will be 17 percent below its calculated capacity. At 1000 ft—typical altitude for much of the United States—the error is 3.6 percent, but there are drying plants in the world above 10,000 ft (3048 m).

To dry a solid that has a negative heat of crystallization requires additional heat. If overlooked, the dryer will be undersized by up to 20 percent or more. Chapters 5 and 6 give calculation and charting methods to determine the effect of this property on airflow and heat load. The effect is proportionately less for lower-moisture feeds, but it should never be ignored. Potassium carbonate, for instance, has a heat of crystallization of − 85.7 Btu/lb, and it raises the airflow by up to 4.4 percent, depending on feed and product moisture contents. Magnesium chloride's value is − 686 Btu/lb; it raises the airflow by as much as 35 percent.

Feed and evaporation rates.   As noted in Chap. 7, the measurement of feed rates is not given a high priority at most plants. This is in spite of its importance in determining the evaporation rate, which is essential in evaluating dryer performance. The most direct route to evaporation rate is the difference between feed and product rates. Figuring it from the flowrate, temperature, and moisture in the air is generally less accurate.

Wet solids feed rates usually can be determined only by weighing. For solutions, rotameters are the most commonly used devices, but slurries require magnetic flow meters or other types. Some firms determine the feed rate by using weigh tanks with load cells.

The most important condition over which there is plant control is the feed moisture content (or solids content). Its magnitude and how it varies has a major effect on the evaporation rate and production rates. It may also have an influence on product quality and on the transport of solids through the dryer. If there is any doubt about the value of feed moisture, it should be measured on a regular basis.

Leaks.   Leaks are a major challenge in testing, commissioning, and operating, as well as troubleshooting. Calculating leaks from airflow measurements is described in Sec. 7.6. There may be several leaks in a dryer, especially in an old one, but even one leak can cause serious undercapacity.

During a shortage of technicians, a veteran sales engineer was sent to start up a small standard spray dryer. In a few hours he had the new system running well, except that the production rate was well below design. When he telephoned his company office, the chief engineer argued—over his objections—that the access door must have been left open. When finally convinced to go back and check again, he found that the door, while closed, was unlatched. The leakage through about 10 ft of gasket at 2 in WG (0.5 kPa) introduced enough ambient air to cut the capacity by more than one-third.

A sizable in-leakage between the dryer inlet and the blower—the vulnerable area of a drying system—adversely affects productivity and energy use. Even new drying systems have leaks, which get worse with age, often unnoticed in the noise of a plant. Some very old product collectors develop huge leaks, in particular those subject to corrosion. A prime source of leaks are large gasketed flanges, especially those on doors made to fit curved vessels.

Measurements taken on old systems have revealed leaks that made up over 40 percent of the airflow. These were in old bag filter collectors, but leaks in other places can be just as serious. They can diffuse in through insulation, then short-circuit along the inside surface of a duct without much mixing with the drying air. If this occurs in the hot air duct leading to the drying vessel, a temperature recorder with a single probe can give a false reading. The recorder at one 30-year-old plant read 1100°F (593°C) entering the dryer, but a traverse gave an average of only 850°F (454°C). From accurate measurements the technician calculated a leak equal to 34 percent of the hot air.

Layers of hot and cold air can also be created by slip-type joints that allow thermal expansion of the duct between heater and drying vessel. Current practice is to provide a flexible cover, which prevents leakage between the loose fitting parts. In time the covers often split, however, so they should be checked regularly and, when necessary, replaced, preferably with a leakproof accordion type joint.

Leaks can be very elusive, as when cracks develop in equipment or ducts under insulation, or when flanges warp. Listening for them is not a reliable search method. But they can be found successfully, even under insulation, by using smoke, which can be sucked in at several points to triangulate the offending location.

At most plants smoke generators must be used; a glowing ember or cigarette would violate regulations. Some of the sophisticated methods for locating leaks are expensive and inconvenient to use. But these specialized techniques may be needed if airflow measurements or calculations can only be made at or near the system's inlet and outlet.

Even a small leak can cause a major problem. One at the bottom of a cyclone, such as coming through a rotary air lock, hurts collection efficiency without being noticed. A rough test for such a leak can be made with a piece of paper placed under the air lock when the system is running without feed. The paper will be sucked up tight if there is an excessive leak. On the other hand, leaking air locks under bag filters do not usually reduce collection efficiency, but they do cause more product to be held up in the hopper.

Auxiliary airflows. Although many air inflows are undesirable or unavoidable leaks, some are deliberate and may have to be quite large. An example is cooling the product before discharge. Any of those inflows can be estimated by accurately measuring airflows at the significant points and calculating the differences. On the other hand, leaks at the feed entry and product discharge points of rotary and other dryers can be determined by calculating the airflow needed for evaporation and subtracting it from the total airflow. In systems with cyclones often the fastest reasonably accurate method is to determine the airflow rate from their pressure drops.

Ambient air is brought into one type of spray dryer to cool products that are tacky when dry but still warm. Some atomizer drives operating in high temperatures must be cooled with a stream of ambient air, which passes through into the chamber. Because they greatly reduce drying efficiency, deliberate leaks such as these should be kept to an absolute minimum.

12.3.4 Product quality

There are many problems relating to product quality, and each may have several causes. Product properties that are significant in many drying applications include moisture content, bulk density, and average size and distribution of particles. Various other properties are important for specific products—color, taste, flavor, aroma, flowability, and many more.

Previous chapters dealt to some extent with the relationship between properties and operating conditions, but it is a broad subject that would require volumes to cover even superficially. Each of the thousands of products, some in several dried forms, for the most part would have to be treated separately.

12.4 Better Operations

If the production rate has to be increased, several options can be tried. These steps range from simple adjustments to complete system overhauls and are discussed in some detail in Chap. 8. Improvements are possible because most dryers operate at conditions set by limited testing. In order to optimize a drying system, it takes time to analyze product quality and product buildup. Some plants find it more convenient to get data for optimizing in a small test dryer. But the best results are most often obtained in the commercial plant, in its unique setting of equipment, feed material, production rate, and personnel.

The technician should encourage good operations by updating the instruction manuals, at least to the extent of the work done on his or her visit. Plant personnel can be shown how to calculate the evaporation rate from feed and product rates or other measurements. A visit to the plant is also an opportunity for the technician to retrain operators or, if that is not practical, to observe their working procedures and make suggestions. Good operators are interested in learning how the essential variables relate to dryer efficiency.

There are major system modifications that can reduce energy use and improve productivity, but costs are too high for justifying them during troubleshooting work. They may, however, be considered later. Such modifications include feed preheaters, internal heat exchangers (for fluid beds), two-stage drying, and recycling part of the exhaust air. In addition, exchangers to preheat the air with the exhaust reduce energy use, although they provide no productivity benefit. These options are dealt with in Chap. 8.

Chapter

13

Nonaqueous Solvent Systems

13.1 Introduction

Many materials are first produced in organic solvents, but before the drying step the solvents are usually replaced by water. This allows open-cycle drying, using cheap available water and air. In some applications, however, it is not possible or practical to switch to water, so the solvent has to be evaporated in a closed-cycle drying operation. An advantage is the low heat requirement, the major operating cost factor. But there are many disadvantages, not the least of which are the hazards in handling these solvents and the design complexities they impose. To a large degree, drying from solvents other than water is another realm.

A solvent dispersed in air and within a certain flammability range can explode if ignited. Data on lower explosive limits (LEL) and upper explosive limits (UEL) in air by volume are given in App. B. It is a general rule, in the interest of safety, to avoid any occurrence of a vapor-air mixture that is over one-quarter or, at the most, one-third of the LEL. A mixture of solvent in air above the UEL is not ordinarily considered to be safe. In most cases such a mixture could easily be diluted with more air, and thus reach the explosive range. The weight ratio of solvent vapor to air S_r that is safely below the LEL is found from

$$S_r = \frac{\text{LEL } M_v}{S_f(100 - \text{LEL})M_a} \tag{13.1}$$

where M_v and M_a are the molecular weights of solvent and air, and S_f is the safety factor. A safety factor of 3 is recommended as a minimum.

Drying gases. The gas used as the drying medium must not cause any problems with the solvent (or the product). Air creates an explosive mixture with most solvents; thus a nonreacting gas is required, often referred to as an *inert* gas. It is usually nitrogen, and its properties are only slightly different from those of air. Carbon dioxide, methane, and other gases have been used occasionally.

The closed system required for solvents is also useful for handling products or solvents that are toxic or have other hazards. The choice of drying gas depends on the kind of hazard; in a few cases it can be air. Some systems are built to recover the solvent, and the dry substance is a byproduct.

Solvents. Water is in many ways the best, as well as the best known, solvent, but in this chapter the term *solvent* is used for any non-aqueous solvents. The term *moisture*, on the other hand, is used for solvent vapor and liquid, just as it is for water. Virtually all of the nonaqueous solvents are organic, and most are flammable, requiring special designs for safe operation. Recovering them after the drying step is an economic necessity, normally achieved by condensing. Adsorbing is an alternative for small systems. The properties of some alcohols, aromatics, hydrocarbons, chlorinated hydrocarbons, and ketones that are commonly used in drying systems are listed in App. B.

In spite of the problems, it is sometimes cost-effective to evaporate the solvent, even if it could be replaced by water. The choice of solvent is generally dictated by the prior process—reacting or mixing in most cases. But if the solvent can be chosen to suit the drying operation, it should have a low latent heat of vaporization, have no hazards that cannot be managed, and cause no drying problems.

Often a mixture of solvents is used—normally miscible or partially miscible—and they have complex relationships. Calculating these for drying is complex, and information is sparse, even for mixtures of just two solvents. The effect of the presence of inert gas on drying and condensing systems adds still another dimension.

When the pair is water and a solvent that is miscible in it, the one with the lower boiling point is removed preferentially. Thus if their percentages in the feed are equal and water has the higher boiling point, the final product moisture will be mostly water.

Just as in air-water systems, the efficiency of solvent drying is improved by higher temperatures of the inlet gas and feed, lower outlet gas temperature, and lower feed moisture. The quantity of drying gas determines equipment size and cost, while the approach to saturation influences product moisture.

Drying operation.   Nearly all the common direct and indirect dryers are used successfully on solvent drying applications. Evaporation, at least initially, is much faster than in an air-water system because of the low latent heat of vaporization. This creates some unique problems, as when solutions are sprayed and the material dries before drops can form.

In spray drying, the rapid evaporation often forms strings or filaments that, when taken from the dryer, look like cotton candy. A similar condition—also most common in the evaporation of solvents—is the clustering of discrete particles held together by static charges. These formations have extremely low loose bulk density and are almost never acceptable. The primary ways to avoid them are to slow the rate of drying or speed the formation of drops—dilute the solution, lower the inlet drying temperature, or produce larger droplets. Alternatively, a cool gas can be introduced to contact the droplets before they start drying, and thus allow them to form properly.

Product moisture.   At the outlet of a solvent dryer the vapor content in the gas is 10 to 60 times greater than that for water vapor in air at the same conditions. This lower diffusional driving force reduces the rate of evaporation. In addition, a solid may have a special affinity for the solvent. Thus it is often more difficult to reach a low solvent content in a product without higher outlet temperature, longer exposure time, or secondary drying. Even traces of solvent in a product may be unacceptable. Retained water, in contrast, generally either is harmless or serves a useful purpose.

13.2   Closed-System Requirements

Most solvent systems must meet the following requirements:

1. Recover the solvent for safety reasons, to meet environmental laws, and because of its value

2. Prevent an explosive mixture of solvent vapor with air by using an inert-gas drying medium

3. Recycle the drying gas because of its value and to allow recovery of all of the solvent

4. Prevent the escape of vapor and creation of sparks or other ignition sources in or near the system

5. Permit start-up from an air-bound condition, adding inert gas and bleeding off the air until the oxygen content is at an acceptably low level

6. Dry to a low moisture content to avoid any harmful results from the solvent in the product

7. Suit the unique properties of the solvent and drying gas, providing the necessary precautions for toxicity and other hazards

A completely closed system is similar to the one shown in Fig. 2.4, except that the heater would be an indirect type and the exhaust–supply air heat exchanger would not likely be included. Vapor concentration in the recycled gas reaches a level that balances solvent input and output. The amount of bleed would be much smaller, except for start-up. The scrubber serves as both backup collector and condenser. It is fed enough of the same solvent as in the feed to cool the gas sufficiently to condense vapor. Alternatively, a tubular type of condenser could be used, but the low-pressure inert gas would result in a very low heat transfer rate.

Design. Calculations for a closed-cycle dryer to handle solvents have to consider the properties of each solvent—unique from all others. The physical design has to be able to separate the product from the gas-vapor mixture, then separate the vapor from the gas and recycle the gas.

Preventing leaks is of prime importance; otherwise the oxygen content will build up. Leaks, either in or out, can create explosive mixtures. Outward leaks are to a degree self-indicating and less dangerous, however, so systems are usually built to operate at a slight positive pressure. It is essential that the oxygen level in the system be analyzed on a continuous basis, with special care given during start-up.

The hazards of most organic solvents mandate extreme caution in handling and in equipment design. In the United States, government regulations place restrictions on the use of some solvents and on the pollution control of most.

A special problem in liquid solvent lines is the buildup of static charges, which, if released, could ignite an explosive mixture. Ground cables on flanges around gaskets reduce the chances for static sparks, as do nonsparking tools and all moving parts, such as fan impellers. To minimize leaks, good flange and shaft seal designs are essential. Any equipment that requires gas cooling or purging, such as bag filter collectors, should use nitrogen or some other inert gas. Any gas introduced into the system has to be continuously bled out.

Costs. The required equipment volume is much lower for a solvent system than for a water system. The latent heat of vaporization of some solvents is only 20 percent that of water, and alcohols are at the most only twice that. The high ratio of solvent to gas, compared to a water vapor–air system, provides a higher heat capacity of the mixture. But that benefit is offset for some applications by the difficulty in reaching the desired product moisture.

Nearly always the net result is much lower energy costs and a smaller drying vessel. Costs for a total system and for overall operating costs, however, may not be lower. Several extra items that raise costs include closing the loop, providing explosion venting, and removing the solvent from the gas. In addition, indirect heaters are required, and this puts an upper limit on the maximum drying gas temperature of about 600 to 800°F (316 to 427°C).

Certain peripheral items also add to the cost of drying from a solvent. One is the need to rework the recovered solvent by distillation or other processes. Another is the special building features, such as explosion venting and venting of drains and other elements. In addition, various components, such as electrical equipment, have to meet explosion regulations, and insurance costs are higher.

Superheated vapor systems. It is sometimes possible to use the superheated vapor of the solvent itself as the drying medium and eliminate the inert gas. When feasible, it requires much smaller equipment volume and cost because of the high solvent molecular weight. Solvent recovery is much simpler when condensing without the inert gas, except at start-up. A serious drawback, however, is the possibility of condensation in the dryer, unless the outlet drying temperature is kept well above the solvent boiling point. Also, for any specific product it may be difficult to reach the desired final moisture because of a lower diffusional driving force.

13.3 Calculation Methods

There are several ways to perform the basic dryer calculations of gas flow rate, heat and condensing loads, and saturation condition. They include the following:

1. Manual calculations

2. A computer program based on the manual procedure

3. A psychrometric chart, such as Fig. 13.1, from which the needed data can be taken

4. One or more nomographs, such as Figs. 13.2 and 13.3, for estimating the gas flow rate and the heat load

All these methods use the solvent properties, which often are not easily located. Properties are different for each solvent but, as Figs. 13.2 and 13.3 show, results are similar for some closely related homologs. The necessary properties of the most common solvents, gleaned from various sources, are given in App. B, as are upper and lower explosive limits. The hazards, complexities, and uncertainties of solvent drying systems put a premium on reliable data and methods for estimating and calculating. Thus generous safety factors of 10 to 20 percent are recommended.

13.3.1 Calculation by equations

Introduction.   This procedure, like the one for the air–water vapor calculation of Chap. 5, finds the flow rate of the drying gas and the heat load. Other items calculated are the cooling load of the condenser and the saturation condition at the condenser outlet. Items not repeated because they are similar to those of Chap. 5 are operating pressure from the elevation and feed and product flow rates. Vessel sizing is substantially the same as that covered in Sec. 5.2.6, but compared to water, a longer exposure time, a higher outlet temperature, or secondary drying may be needed.

Calculations differ in several ways for solvent operation as compared to those for an air–water vapor system. It is a reasonable assumption that the collection efficiency is 100 percent and that there are no leaks or auxiliary gas flows. Without the extra gas flows, only stations 1, 2, and 3 are needed—condenser outlet and dryer inlet and outlet. In this procedure the gas flow is found only for station 3, but the others could be added easily.

Another difference is the use of latent and specific heats of the solvent to compute enthalpies and the saturation condition at the condenser outlet. This is less accurate than using equations derived from known data, as for the air–water vapor system, but few such data are available for solvents.

Thus the vapor enthalpy is found by summing the sensible and latent heats. These are liquid from 32°F (0°C) to the boiling point, evaporation at the boiling point, and vapor sensible heat from the boiling point to the desired temperature. (The last item is negative if the outlet temperature is below the boiling point.) A more accurate method would be to use a lower evaporation temperature, one that is closer to the average drying temperature. But that point is not well known and requires a calculation of latent heat. In any case, the accuracy gained is probably not worth the effort.

Nomenclature and units.   Variables used in Chap. 5 that apply here have the same names. The data arrays for solvent vapor pressures and their corresponding temperatures are from the Stull data reported in Perry and Green (1984). They are required in the procedure and given in App. B. The basis of enthalpies, moistures, and specific volumes is a weight unit of dry gas (DG), just as a unit of dry air (DA) is the basis in air-water calculations.

A	Coefficient for vapor pressure equation
B	Coefficient for vapor pressure equation
C_g	Specific heat of gas, Btu/(lb · °F)
C_q	Specific heat of liquid, Btu/(lb · °F)
C_s	Specific heat of solid, Btu/(lb · °F)
C_v	Specific heat of vapor, Btu/(lb · °F)
E_p	Ratio of evaporation to product rates
E_v	Evaporation rate, lb/h
F_3	Volumetric gas flow rate at dryer outlet, ft^3/min
H	Gas-vapor enthalpy, Btu/lb DG
H_1, H_2, H_3	Gas-vapor enthalpies, Btu/lb DG
H_5	Initial value of H_3, Btu/lb DG
H_c	Heat of crystallization, Btu/lb product
H_d	Heat effect from product solids, Btu/lb DG
H_e	Heat effect from feed liquid, Btu/lb DG
H_m	Heat effect from product moisture, Btu/lb DG
H_h	Heat effect from heat of crystallization, Btu/lb DG
H_f	Total heat effect, Btu/lb DG
H_r	Heat loss, % of H_2
K_c	Constant, = 2.3026
L_h	Latent heat of vaporization, Btu/lb solvent
M	Moisture content (absolute humidity), lb/lb DG
M_1, M_2, M_3	Moisture contents (absolute humidity) lb/lb DG
M_g	Molecular weight of gas
M_v	Molecular weight of solvent
P_m	Product moisture content, wet basis, %
P_r	Product rate, lb/h
P_t	Pressure, lb/in^2
Q_c	Cooling load for condenser, Btu/h
Q_h	Heating load for heater, Btu/h
Q_n	Heat leaving in condensate, Btu/h
S_1, S_2, S_3	Solvent flow rates, lb/h
T	Temperature, °F

T_1, T_2, T_3	Drying gas temperatures, °F
T_b	Solvent boiling point, °F
T_n	Temperature of condensate, °F
T_f	Temperature of feed, °F
T_p	Temperature of product, °F
T_g	Temperature of gas, °F
T_q	Temperature of liquid, °F
T_v	Temperature of vapor, °F
T_k	Temperature, K
T_a, T_z	Temperatures straddling T_1 (for vapor pressure), K
V_3	Humid volume at dryer outlet, ft^3/lb DG
V_a, V_z	Vapor pressures corresponding to T_a and T_z, mm Hg
W	Enthalpy of gas and vapor, Btu/lb DG
W_3	Enthalpy of gas and vapor, Btu/lb DG
W_g	Flow rate of gas, lb/h
Z_a, Z_b, Z_c, Z_d	Coefficients for C_v equation

Bases of procedure.   The calculation procedure is based on the following:

1. Definitions are given under *Nomenclature and units*.

2. Units are in the U.S. customary system, except that data and equations from the literature that are in metric units have not been changed. The equations make the necessary adjustments. See App. A for conversions.

3. The drying gas is nitrogen. Changing to air would have little effect on the results, but if some other gas were used, all the affected properties should be revised.

4. The temperature basis is 32°F (0°C) for both solvent and gas.

5. The weight basis is 1.0 lb DG for enthalpies, humid volumes, and gas moistures.

6. All percentages are assumed to be reduced to fractions before calculations begin.

7. No safety factor has been included. Adding 10 to 20 percent is recommended for critical items not adversely affected by increased size (such as cyclones).

8. Solvent property data needed for the calculation are given in App. B, Table B.6, except for vapor pressures, which can be found in Perry and Green (1984).

Equations for calculating gas flow and heat load. For pressure P_t and the feed, product, and evaporation rates see Eqs. (5.7)–(5.16) in Chap. 5. For those equations it is assumed that there is no powder loss, so $E_t = 1.0$. Thus $D_s = S_p$ and $P_1 = 0.0$. The ratio of evaporation to product is

$$E_p = \frac{E_v}{P_r} \tag{13.2}$$

A modified Antoine equation is used to interpolate saturated moisture for station 1 gas out of the condenser. Five-place accuracy is required. Convert T_1 to degrees Celsius, then select from the temperature and vapor pressure data the two temperatures that straddle T_1. Their corresponding vapor pressures are V_a and V_z. Using Eq. (13.3) convert the °C values of T_1 and the two data temperatures that straddle T_1 to kelvins. These become T_k, T_a, and T_z:

$$\text{Kelvin } T = T + 273.15 \tag{13.3}$$

$$B = \frac{\log(V_z/V_a)}{1/T_z - 1/T_a} \tag{13.4}$$

$$A = \log(V_a) - \frac{B}{T_a} \tag{13.5}$$

$$V_p = \exp\left[K_c\left(A + \frac{B}{T_k}\right)\right] \tag{13.6}$$

The moisture ratio is saturated at station 1, and though not saturated at station 2, it has the same value in lb/lb DG,

$$M_2 = M_1 = \frac{V_p M_v}{(P_t - V_p)M_g} \tag{13.7}$$

Equations (13.27)–(13.32) form a separate calculation segment, named subroutine X. It is used to calculate specific heats and enthalpies for each of the three stations:

$$T = T_1; M = M_1: \quad \text{Using subroutine } X \text{ calculate } H_1 \tag{13.8}$$

$$T = T_2; M = M_2: \quad \text{Using subroutine } X \text{ calculate } H_2 \tag{13.9}$$

$$T = T_3: \quad \text{Using subroutine } X \text{ calculate } W_3 \tag{13.10}$$

W_3 will be needed for calculating the unknown M_3. H_5 is the initial value of H_3, found by reducing H_2 by the heat loss H_r. But H_3 has not been adjusted for other heat effects,

$$H_3 = H_5 = H_2(1 - H_r) \tag{13.11}$$

M_3 is based on H_3, which changes after adjusting for the heat effects. Thus iteration is required of Eqs. (13.12)–(13.18) until H_3 remains constant,

$$M_3 = \frac{H_3 - (T_3 - T_t)C_g}{W_3} \tag{13.12}$$

$$H_e = (M_3 - M_2)(T_f - T_p)\, C_q / E_p \tag{13.13}$$

$$H_d = \frac{(M_3 - M_2)(1 - P_m)\, C_8(T_f - 32)}{E_p} \tag{13.14}$$

$$H_m = \frac{(M_3 - M_2)P_m C_q(T_p - 32)}{E_p} \tag{13.15}$$

$$H_h = \frac{(M_3 - M_2)(1 - P_m)H_c}{E_p} \tag{13.16}$$

$$H_f = H_e - H_d - H_m + H_h \tag{13.17}$$

$$H_3 = H_5 + H_f \tag{13.18}$$

When H_3 is within about 0.2 of its previous value, the convergence should be acceptable, and the solvent and gas flow rates and the condensing and heat loads can be figured:

$$W_g = \frac{E_v}{M_3 - M_2} \tag{13.19}$$

$$S_3 = W_g M_3 \tag{13.20}$$

$$S_1 = S_2 = W_g M_2 \tag{13.21}$$

$$V_3 = 555(T_3 + 460)\,\frac{S_3/M_v + W_g/M_g}{P_t W_g} \tag{13.22}$$

$$F_3 = \frac{W_g V_3}{60} \tag{13.23}$$

$$Q_c = E_v\left[(T_3 - T_b)\, C_v + L_h + (T_b - T_n)C_q\right] + S_1(T_3 - T_1)C_v$$
$$+ W_g(T_3 - T_1)C_g \tag{13.24}$$

$$Q_h = W_g(T_2 - T_1)C_g + S_2(T_2 - T_1)C_v \tag{13.25}$$

$$Q_n = E_v(T_n - 32)C_q \tag{13.26}$$

Subroutine X. This subroutine calculates specific heats and enthalpies at the assigned values of T and M, after converting T to °C or K, as required,

$$T_g = \frac{(T + 32)/2 + 460}{1.8} \tag{13.27}$$

$$C_g = \frac{6.83 + 0.0009T_g - 12{,}000/T_g^2}{M_g} \tag{13.28}$$

Table B.6 in App. B lists the liquid specific heat C_q for each solvent as a constant. Values of C_q were calculated from specific heat data, using the average temperature between the reference temperature, 0°C, and the boiling point.

$$T_v = \frac{(T + T_b)/2 + 460}{1.8} \tag{13.29}$$

$$C_v = \left(Z_a + \frac{Z_b}{100} T_v + \frac{Z_c}{10^6} T_v^2 + \frac{Z_d}{10^9} T_v^3\right) M_v^{-1} \tag{13.30}$$

$$W = (T_b - 32)C_q + L_h + (T - T_b)C_v \tag{13.31}$$

$$H = (T - 32)C_g + MW \tag{13.32}$$

This terminates subroutine X.

The dew point can be calculated by selecting a trial temperature T and using Eqs. (13.3)–(13.6) to get the saturation moisture M_s. Then iterate down from T_b until $M_s = M_3$. The saturation temperature and the adiabatic saturation ratio can be calculated using the saturation moisture equations and Eqs. (13.27)–(13.32) for the corresponding enthalpies. Wet-bulb temperatures can be figured by a method similar to the one in Chap. 5, Eqs. (5.80) and (5.81). The result will be uncertain, however, because the method is based on a straight-line relationship that applies only to water vapor in air.

13.3.2 Calculation by psychrometric charts

The equations in Sec. 13.3.1 can be modified, in the same manner as those in Chap. 5, to allow computing data for solvent psychrometric charts. At similar temperatures, values of absolute moisture are much higher for solvent systems, making the enthalpies higher in spite of lower latent heats of vaporization.

A chart for hexane is shown in Fig. 13.1, plotted in the same manner as the air–water vapor charts of Chap. 6. Drawn on the chart is an example of 60°F condenser outlet (heater inlet), 400°F dryer inlet, and 200°F dryer outlet. Reading values and making the necessary subtrac-

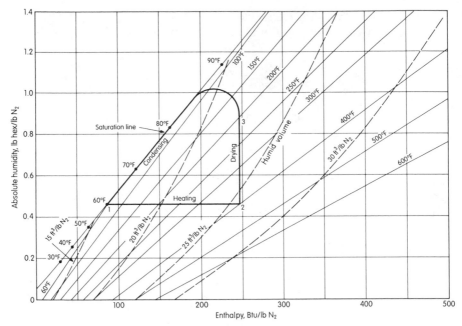

Figure 13.1 Psychrometric chart for hexane in nitrogen.

tions, the moisture difference between stations 2 and 3 is 0.42 lb/lb DG, the humid volume V_3 is approximately 22.0, and the enthalpy difference between stations 2 and 1 is 162 Btu/lb DG. For an evaporation rate E_v of 1000 lb/h, the gas flow rate is

$$W_g = \frac{E_v}{M_3 - M_2} = 2381 \text{ lb/h} \tag{13.33}$$

the volumetric gas flow rate is

$$F_3 = \frac{W_g V_3}{60} = 873 \text{ ft}^3/\text{min} \tag{13.34}$$

and the heat load is

$$Q_h = W_g(H_2 - H_1) = 386{,}000 \text{ Btu/h} \tag{13.35}$$

These values will agree closely with those obtained using the equations in Sec. 13.3.1 if computed on the same basis. But the chart, without special manipulations, does not take into account elevation or heat effects, such as the difference between heat entering in the feed and leaving in the product.

The added operation of condensation removes liquid solvent from the system and cools the gas stream for recycling. The total cooling load Q_c can be calculated from the difference in enthalpies between stations 3 and 1 times the gas flow rate,

$$Q_c = W_g(H_3 - H_1) = 386{,}000 \text{ Btu/h} \qquad (13.36)$$

Heating and cooling requirements are the same, because it is assumed that drying is adiabatic and gas and condensate are cooled to T_1 when, in fact, the condensate leaves the scrubber at a temperature approaching the station 3 wet-bulb temperature. In the example the condensate leaving will be about 80 to 90°F (27 to 32°C), depending on the wet bulb, the efficiency of the scrubber, and the operating conditions of the coolant system. The correct load can be calculated if the temperature of the liquid is known, or if it can be determined by testing.

13.3.3 Nomographs

Figures 13.2 and 13.3 are nomographs that permit estimating very roughly the gas flow rate and the heat load for some common solvents. They are based on a condenser outlet T_1 of 60°F (16°C), selected as a compromise. From a practical standpoint, however, it will be too high for a few solvents and too low for others. Lower values of T_1 increase both main calculation requirements—gas flow rate and heat load—while higher values decrease both, because at lower temperatures the heat capacity of gas and moisture is lower. A change in T_1 of 20°F (11°C) typically affects results in the range of 4 to 15 percent.

Following the path of temperatures T_2, T_3, T_1, shown in Fig. 13.2, yields CE, the ratio of gas flow rate to evaporation rate. Multiplying this by the evaporation rate yields the volumetric gas flow in ft^3/min. In the same manner Fig. 13.3 gives the ratio of heat load per evaporation rate BE, which when multiplied by the evaporation rate yields the heat load in Btu/h.

The example shown on both charts is for hexane at 400°F dryer inlet and 200°F outlet. Results are 1.03 for CE and 425 for BE. When multiplied by an evaporation rate of 1000 lb/h, the results are reasonably close to those obtained from the psychrometric chart example.

Values from Figs. 13.2 and 13.3 include a 10 percent safety factor. They are accurate only to about 15 percent. Values for water are scattered broadly, and are thus shown as a band. The lines for the aromatics are closely bunched, the alcohols and chlorinated hydrocarbons less so, as might be guessed from their molecular structures.

13.4 Testing

No comprehensive coverage of the safety aspects of testing—and their legal, cost, and insurance ramifications—can be given here. An in-depth study is required, giving full consideration to the peculiarities

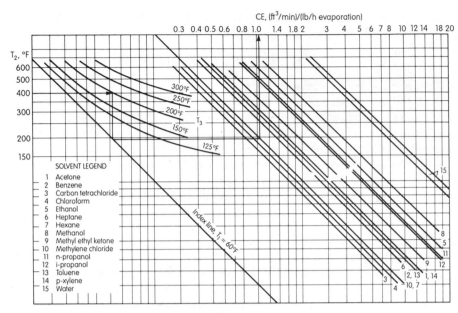

Figure 13.2   Nomograph for gas flow in solvent dryers.

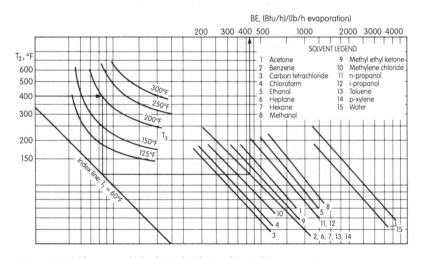

Figure 13.3   Nomograph for heat load in solvent dryers.

of the specific solvents to be handled. The use of nitrogen or some other inert gas as the drying medium does not diminish the need for good safety practices. These include purge at start-up and shutdown, and delaying feeding until the oxygen content is at a safe level for the solvent that is used. Continuous oxygen analysis, grounding against static discharges, and the use of tools and moving equipment parts of nonsparking materials are also essential.

Proper cautions have to be taken to avoid dangerous conditions—fires, explosions, and toxicity—in handling, storing, pumping, drying, packaging, cleanup, and shipping. One simple example on many applications is the need to use some of the solvent when cleaning equipment. Working in solvent fumes in confined spaces requires the right protection. Not only do many solvents present a toxic problem, but especially in the heat of drying, some give off toxic monomers.

Evaluation testing is much the same as for air-water systems, but with strong additional emphasis on safety and the problem of obtaining an acceptably low moisture content in the product. Secondary drying is often needed to reach specifications.

Polyethylene and polypropylene in hexane at 30 to 50 percent feed moisture are frequent solvent applications for indirect dryers. They require a low final product moisture, and tests to demonstrate and evaluate this usually need two-stage drying. A high-speed paddle dryer at its most efficient operation can dry these materials to about 1.0 percent moisture. A low-speed paddle dryer is often a good choice for the second stage. Depending on the product, the exposure time needed is 15 min to 1 h or more to dry down to a final moisture of a few parts per million.

On test dryers it is important that the product receiver be monitored carefully. Incompletely dried product may give off vapor and form an explosive mixture with air in the collector, which often is difficult to seal adequately.

Solvent tests were once run using air in open-system dryers. Such tests were usually only feasibility studies; they are rarely considered or even permitted any longer because of the dangers. For even minimum safety, the feed rate has to be low enough to keep the air-vapor mixture in the dryer at less than one-third of the lower explosive limit. The very low solvent concentration, however, speeds drying abnormally. It can give misleading results, such as by increasing filamentation of sprayed feeds.

Chapter

14

Applications

14.1 Indirect Drying of Sludges and Slimes*

Materials that are difficult to dry need closer scrutiny in the dryer selection process, and they often require special designs, techniques, or auxiliary equipment. A major problem such materials cause is fouling of the dryer heat transfer surfaces. Another, sometimes related, problem is preventing or restricting the escape of vapor from the particles, because during the drying process the material either forms agglomerates or case hardens. Vapor diffuses slowly through enlarged particles or those with a hardened outer zone.

A third problem is material going through a cohesive (plastic) stage, during which it turns with the agitator without moving forward—a condition referred to as *roping*. These difficulties may occur in some of the most common industrial operations, one of which is drying wastewater sludges.

Industrial wastewater. Many industrial processes create wastewater that has to be disposed of in some acceptable manner. Dumping is seldom an option. These wastes vary greatly in chemical composition, consistency, and solids content, so they cannot all be processed in the same way. They are often not accepted in municipal treatment plants, and many firms have to install their own facilities. To meet federal, state, or local laws for ultimate disposal, such as adding to landfills, a dewatering step or a drying step, or both, are usually required.

Plate and frame filter presses, belt filter presses, rotary vacuum filters, and centrifuges are used, but they cannot remove enough water from most sludges. Drying, on the other hand, can take out most of the

---

*William L. Root III, Komline-Sanderson Engineering Corporation.

water, thus making the solids a useful resource—as a soil conditioner (if free of heavy metals) or as a heat source (mixed with a fuel such as powdered coal).

Industrial wastewater is usually treated in two steps. Primary treatment is settling, straining, and possibly skimming. Secondary treatment consists of pumping in air and adding bacteria to speed up the decomposition of organic materials and chemicals. This eliminates most of the substances that consume dissolved oxygen. It also raises the water's oxygen content, making it fit for aquatic life and ultimate recycling into potable water supplies.

Separation steps. After secondary treatment the water and the remaining solids have to be separated before being disposed of, each in its own way. If the water is pure enough, it can be delivered to some water source. Otherwise it will have to be recycled for retreating.

Possible options for separating solids from water, in the order of increasing costliness, are settling, filtering, centrifuging, or possibly some combination of these, and finally drying. Most solids are finely dispersed or dissolved in the water, so settling and screening remove little. Because the solids and the water have an affinity for each other, the sludge is still between 70 and 90 percent liquid after filtering or centrifuging. The spent bacteria and conditioning agents are part of the sludge, and they contribute to its properties. Getting it dry enough to be disposed of is the final step—drying.

Sludge drying problems. Many sludges are sticky or tacky—the particles tend to stick to each other and to other surfaces. This stickiness is often limited to a specific moisture range. It may occur briefly during the drying, usually as the material goes through a plastic stage. How it handles through this period governs the drying techniques to be used and the type of dryer that will succeed.

Dry product can be backmixed into the feed if consistency is a severe problem. By raising its solids content the difficulty is bypassed. Costs are increased, however, by the addition of a backmixer and transfer equipment and by increasing the dryer size. There is only a small offsetting benefit from heat in the recycled product.

Textile sludge plant. Waste treatment at textile printing operations is fairly typical of these operations. Textiles from the weaver contain sizing and oil, which have to be washed out and rinsed. The wash water, which also contains short fibers, is held in a pond for several days, while air is pumped in vigorously and bacteria are added. This reduces the biological oxygen demand (BOD) and the chemical oxygen demand (COD) to acceptable levels. The liquid, conditioned with a polymer, is pumped to a continuous-belt filter press that lowers the moisture content to 85 percent to prepare the activated, biological sludge for drying.

This sludge is sticky at a moisture content of 85 percent and even a little lower. At the start of drying it balls up, reducing the surface available for the escape of vapor. It also has a strong tendency to rope. Conventional disc dryers are very difficult to operate continuously, because their surfaces foul and the solids do not transport properly in the dryer. On the other hand, dual-shafted units with segmented disc designs that scrub—rather than smear—the solids against the surfaces keep them relatively clean. The reduced fouling maintains better contact between solids and metal and improves heat transfer.

Dryer design.   A textile sludge plant has a large paddle dryer with segmented discs that give this scrubbing action. The dual-agitator shafts are 200 in (5 m) long, with segmented discs 34 in (86 cm) in diameter that are almost semicircular and wedge-shaped. Each disc intermeshes with, and rotates counter to, discs on its companion shaft. This produces a kneading action that, together with the drying, breaks up lumps. Solids transport is induced by the angle at which the discs are mounted on the shaft and by the clips on their periphery.

The agitators, turning at 15 r/min, are about 90 percent covered by the solids. The product is dammed up by a weir at the discharge end, which compensates for the shrinkage ratio of 3:1. Because of the shrinkage, the exposure time is not especially meaningful, but it is several hours.

Feed from the filter press at ambient temperature is fed by a cylindrical choke screw conveyor directly to the dryer with no backmixing. The discharge conveyor is also of the choke type, and the small inleakage of air it allows has very little effect on the drying. The jacket and agitators of the dryer are both steam-heated, but the agitators contribute most of the heating in both surface area and heat transfer rate. The average rate is 27.4 Btu/(h · ft^2 · °F) [155.6 W/(m^2 · K)], which is high for a sludge, but lower than for most other materials in this type of dryer.

The particle size of the final product is 96 percent minus 60 mesh, and the rest are balls averaging ¾ in (1.9 cm) in diameter. The product, at well under 10-percent moisture, is mixed with pulverized coal and fed to the burner in the plant's boiler.

The system has programmable control and is attended on the day shift, but runs unattended the rest of the time. Operation is simple enough and suitably automatic that if it shuts down, it can be restarted easily. One evening the system shut down after a belt on the filter press became misaligned. The night superintendent at the printing plant was alerted by the shutdown alarm signal. When he reached the sludge plant, although he was not very familiar with its operation, he was able to straighten the belt and restart the system with little delay.

Operating data.   Figure 14.1 presents data from a drying test on the textile sludge. Moisture content and heat transfer rate are plotted against dryer length. The percentages of dryer length (and heating surface) are: about 15 percent to reach the boiling point, 26 percent at constant rate, and 59 percent at falling rate. Table 14.1 lists data on the drying of activated biological sludge.

Other problem materials.   Various substances, including other sludges, can take advantage of these drying techniques. Some cannot, however, because consistency problems occur later in the drying. Wood pulps coated with starch, for example, might begin to exhibit a thick, plastic condition at about 40 percent moisture. Materials like that have to be discharged from the dryer before the problems are reached, and other methods used to get them drier.

Some dried sludges, such as those from whey and beer processing, can be sold for fertilizer. Some can be sent to landfills, but sewage sludges and others that contain any harmful contaminants most likely have to be burned. If too wet, they will not support combustion and can cause boiler problems, but they do not have to be bone-dry. The minimum ratio of fuel to solids is determined by the solid's combustion properties and moisture content.

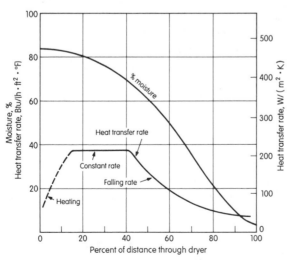

Figure 14.1   Moisture in solids and heat transfer rate for sludge drying test.

TABLE 14.1   Data for Textile Sludge Drying

	U.S. customary units	SI units
Feed rate	2325 lb/h	1055 kg/h
Feed moisture	85%	85%
Feed temperature	70°F	21°C
Product rate	388 lb/h	176 kg/h
Product temperature	225–250°F	107–121°C
Product moisture	10%	10%
Shrinkage ratio	3:1	3:1
Particle size, through 60 mesh	96%	96%
Dryer volume	187.1 ft^3	5.30 m^3
Agitator diameter	34.0 in	86.4 cm
Agitator length	200.0 in	5.08 m
Agitator speed	15 r/min	15 r/min
Heat load	2,204,000 Btu/h	645.9 kW
Agitator surface	479 ft^2	44.5 m^2
Trough jacket surface	159 ft^2	14.8 m^2
Total surface	638 ft^2	59.3 m^2
Heating medium	Steam	Steam
Temperature	338°F	170°C
Pressure, gage	100 lb/in^2	689 kPa
Steam consumption	2503 lb/h	1135 kg/h
Power	60 hp	44.8 W

Drying slime. Copper anode slime is an example of an especially difficult material, but it can be dried using the right techniques. It, too, goes through a sticky phase, which is present in the feed, but it continues well into the drying cycle. Backmixing a high percentage of dry product into the feed before drying gets the sludge out of the moisture range where fouling occurs.

A second problem is the slime's tendency to ball up and for the balls to case harden, retarding the release of vapor from the start. To help overcome both the sticking and the balling problems, a high-speed pug mill can be installed for backmixing. The dual-blade design of the mill first kneads dry powder into the pasty mass of slime, then it cuts and dices the mass.

With all this feed preparation, there is still a tendency of the material to ball up and adhere to metal surfaces during the early stage of drying in conventional disc dryers. But intermeshing wedge-shaped discs of a paddle dryer are self-cleaning to an extent, and they reduce the sticking effects.

Thixotropic feeds. A quite different problem is found in thixotropic feeds. These are fed to the dryer as seemingly dry pastes, but turn liquid from the shearing forces of agitation. Proper agitator design and temperature control prevent the liquid from flowing down the dryer and out the discharge. A short distance from the feed inlet, partly dry material is made to form a dam of solids that holds back the liquid. The upstream end of this dam continuously builds up in place, while the downstream end dries and breaks off into lumps. Drying then proceeds normally, with particles discharged at about 30 to 60 mesh.

14.2   Spray Drying Electronic Ceramics*

Many classes of materials use spray drying, but none to better effect than the various subdivisions of ceramics—white ware, sanitary ware, tiles, clays, and electronic ceramics. Each is a mixture of metal oxides with unique characteristics and methods of treatment. Among all of them, electronic ceramics have the most complex processing techniques and the highest value added. Some common types are aluminas, ferrites, forsterites, silicon carbides, steatites, and titanates.

Few classes of materials, if any, make more exacting demands on the drying process than do electronic ceramics. But drying is only one of many steps needed to make the final parts. There are 14 or more process steps in the usual operation. The main items of equipment typically are calciner, ball mill, blunger, dryer, classifiers, press, and kiln. (A blunger is a tall small-diameter heavy-duty vessel for mixing intensively the high-density ceramics and additives in water.)

Reasons for spray drying.   Considering only the equipment, spray dryers are more expensive than others, such as the traditional tray dryers. But they usually make several steps unnecessary—filtering, crushing, rewetting, and classifying. This more than offsets the higher cost of the dryer and, more important, the overall results are a better product and fewer rejects. The basic steps around the drying are milling the material to submicron size, mixing in additives, making uniform dry spheres, pressing to form the *green* pieces, and firing to the final condition.

The mechanical and electrical properties of the final pieces are greatly influenced by changes in the material that occur during milling, pressing, and firing, but especially in drying. Traditional drying methods make very irregular shapes and sizes that yield a large fraction of fines and a high rate of rejection of the final pieces.

Additives.   To perform properly in drying, pressing, and firing, the metal oxides must be finely ground and have certain materials added. These additives are dissolved in water and uniformly surround the in-

---

*Harman D. DuMont, DuMont Drying Consultants, Inc.

soluble solids. A blunger agitates this slip (slurry) intensively. Virtually all additives are organic; an example is the commonly used binder polyvinyl alcohol. Formulations are closely guarded, and users will seldom disclose the additives or their proportions.

The three types of additives commonly used in electronic ceramics total less than 2 percent of the dry solids in the slip, but they serve vital functions during and after drying.

1. Deflocculents foster higher solids content—65 percent is typical—and that increases particle density.

2. Binders are adhesives. (a) They bind into strong agglomerates the submicron particles in each drop of slip as it dries; (b) they bind together the dried particles when pressed to give them the needed green strength (cohesiveness in the cold-formed powder after pressing); and (c) they may serve as lubricant if of a soft type.

3. Lubricants serve several purposes. (a) They minimize the force needed in the press; (b) they allow the parts to be easily ejected from the dies without breakage; and (c) they reduce wear on the dies and thus extend their life.

As the water evaporates in spray drying, the additives are coated on the particle surfaces. The even coating minimizes the amount needed; an uneven coating may partly burn out in drying. At high outlet air temperatures and long exposure times—needed for low product moisture content—scorching can result. The mixed airflow of tower dryers gives relatively high product temperatures, limiting the use of organics, but not to any serious extent.

Some additives, depending on their concentration in the slip, tend to form films, which restricts their use. But the concentration of additives is low, so most of them have no adverse effect on drying.

Product properties. For final parts to be consistently acceptable, the dried particles have to meet property specifications within close tolerances. The particles undergo major changes before, during, and after drying. They are ball milled to less than 1 micron (μm), forming the dryer feed material, called slip. When spray-dried with various organic additives, they become spheres with average diameters of 80 to 180 μm. After drying they are pressed into pieces of accurate size and weight and, in some cases, machined to more precise dimensions. Then they are fired, where proper allowance has to be made for shrinkage.

The dried ceramic particles, and the parts made from them by pressing and firing, need these characteristics in most applications.

1. High green strength in the pressed particles

2. Identical weight for every part

3. Minimum shrinkage in firing

4. Minimum voids after firing

5. Minimum residue from additives after firing

For the parts to meet specifications, the dried particles must have the following properties and capabilities.

1. Minimum use of additives

2. Uniform coating of additives

3. Specified size and size range of spheres

4. Exact bulk density

5. Minimum fraction of hollow particles

6. Free-flowing so dies are filled completely

7. Low product moisture content, usually less than 0.3 percent

To produce parts with exact electrical, dimensional, and mechanical properties, dried ceramic powders have to meet other high standards, such as for bulk density. Both density and flow are influenced by the shape, size, and size range of the dried particles.

Particle size and size range are governed by the size of the drops that can dry in the time allowed. Figure 14.2 shows the size range of powders collected from the chambers of three dryers. Normally such data would form straight lines. From ceramic dryers, however, the

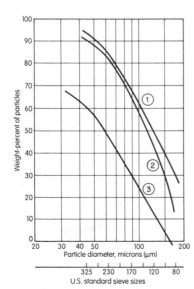

Figure 14.2 Plot of ceramic particle sizes from three dryers. 1—16-ft-diameter chamber with spinning disc; 2—4½-ft- diameter tower with nozzle; 3—2½-ft-diameter tower with nozzle.

fines are taken out separately in the airstream and not included in the usable product. This accounts for the tailing off at the top of the curves. Furthermore, the tower with a 4.5-ft (1.4-m) diameter is often too small to properly dry particles at the high end of its normal range; thus the sag at the bottom (largest sizes) of its curve.

Some ceramics require more strict particle specifications than others. Alumina for spark plugs, like many others, must flow readily into a die and fill it with no voids. In addition, the intricate green part must be ejected without breaking from the die after pressing. It also has to be tough enough to withstand machining on a lathe before firing.

The usual products from spray dryers, as described in Chap. 4, are somewhat spherical, often hollow particles. Most ceramics, on the other hand, dry into nearly perfect spheres, which accounts for their ability to flow, and they are solid. Ceramic particles that are nonspherical or hollow may result when the feed solids content is too low. Such particles are generally unsatisfactory in ultimate use and require that drying conditions be changed or the formulation adjusted.

For every die to receive the right weight of powder, the powder bulk density has to be within close limits, and it usually has to be relatively high. The density is greater for a higher solids content in the slip. It is also greater if the range of particle sizes is wide, allowing smaller particles to fit into the spaces between larger ones.

Often a compromise has to be drawn between a nonuniform particle size for high density and uniform particle size for good flow. Using the tower chamber design, about 80 percent of the dried material is coarse, taken out of the bottom of the chamber. The balance are fines, carried out in the air and then collected. When nonuniform particles are needed for high density, fines are mixed in with the coarse. When uniform particles are needed for better flow, the fines are reworked.

Testing.   As with other products, tests are needed to find the best drying temperatures at clean operating conditions. But the right slip formulation also has to be found. A ceramic engineer or technician is generally needed to evaluate the dried product and to adjust the formulation. Samples of powder often have to be pressed, fired, and tested in actual use before a final judgment can be made. Thus test programs take longer than for other materials.

When testing, the tendency is to aim for ever larger particles, sometimes too large for the test unit. The results, especially for short runs, can be puzzling. After changing the nozzle or its pressure to get larger drops, the operator may find that the dried particles are of the same size range as before the change, and the product has none of the expected large particles. Even though large droplets are sprayed out, they fail to dry completely if the chamber diameter is too small; they adhere to the walls instead. To get the desired size range in commercial units, it has been necessary, on occasion, to replace a chamber with one that is 25 to 35 percent larger while keeping the same airflow rate.

Towers, chambers, and atomization. Ceramics should be tested on a sufficiently large dryer, so that results can be transferred to commercial design. The particle size and size range are affected by the type of atomization, which changes from nozzle to spinning disc when chambers reach a certain size. A two-fluid nozzle is standard in smaller towers. But for larger towers a single-fluid nozzle is preferred, provided the slip is not too abrasive or tends to plug and the product does not need a high density. The crossover point from nozzle to centrifugal atomizer is not sharply defined. For some applications a 10-ft (3.0-m)-diameter tower works well, for others it has to be a 16-ft (5.5-m) or larger conical chamber.

The need for large spherical particles of powder, usually at low production rates, influences chamber design. Large drops need more time to dry. Otherwise they reach the wall while still wet and stick there. To achieve a long exposure time in a small diameter requires long travel for the drops. For low capacities, the best designs are tall towers, similar to the one shown in Fig. 14.3.

The nozzle mounted in the cone throws the spray upward. Thus the drops travel a little less than twice the straight-side height of the tower. The coarse fraction drops out at the bottom, and the fines are carried off in the air. In most cases the practical upper limit of tower design with two-fluid nozzles is 8-ft (2.4-m) diameter, for which the maximum production rate is about 400 lb/h (181 kg/h). Although few electronic ceramics have production rates that can justify large chambers, some alumina dryers are among the largest built for any material. In general, larger diameters, whether towers or conical chambers, allow the production of larger particles, as illustrated in Fig. 14.2.

Above 400 lb/h (181 kg/h) it is more practical to atomize with a centrifugal wheel, for which a large chamber is required. At a diameter of 16 ft (4.9 m) the radial travel from wheel to wall is long enough to dry the large drops. Smaller chambers are inadequate, as evidenced by a 14-ft (4.3-m) dryer that has run for years on alumina, but with frequent problems of wall buildup.

For most ceramics the operating conditions are quite similar. Inlet and outlet air temperatures are usually about 600 and 300°F (316 and 149°C), respectively. This has made possible a series of standard designs—all resembling Fig. 14.3—the largest having an 8-ft (2.4-m) diameter. The height-to-diameter ratios for the towers range between

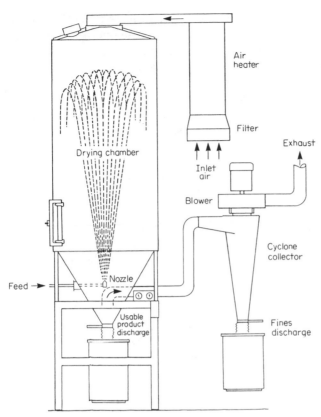

Figure 14.3 Spray dryer for ceramics.

2.1 and 3.3, and are determined largely by field experience in avoiding wet dryer walls and roof. Calculations of airflow and heat load for tower dryers have to consider the heat leaving in the solids, the high heat loss from the tower, and the effect of any compressed air on drying air temperature.

Conventional spray dryers cannot easily be converted for ceramic duty. To provide long enough exposure times, airflow rates for ceramics compared to most other products are only 35 to 50 percent as high. For either tower or conical chamber design, the ducts would probably have to be changed, as might some of the system components. In a tower design it is easy to substitute a ceramic-type nozzle, but it has to be installed in the cone spraying upward. Converting from one atomization type to the other is not a practical option, either. In a tower the horizontal spray of a wheel would not have enough distance in which to dry, while in a conical chamber the spray of a single nozzle would not adequately mix into the airstream.

Nozzle atomizers. A major problem in atomizing ceramics is the extreme abrasiveness of the slips. Wear of the nozzle opening allows an increase in both flow rate and drop size, a condition that can only get worse. Fortunately large particles, and thus coarse drop sizes, are nearly always desired. Thus the wear is less severe than if fine atomizing—at higher pressure—were needed.

For spray drying abrasive feeds, nozzles and centrifugal wheels that resist wear have been developed. One single-fluid nozzle uses a hard-surface material for the opening. A two-fluid nozzle of the external mixing type, as shown in Fig. 14.4, atomizes the stream of liquid after it leaves the nozzle. Atomizing energy is supplied by compressed air or some other gas at pressures of 30 to 80 lb/in^2 (207 to 552 kPa).

Using the nozzle in Fig. 14.4, the drop size is controlled by regulating the feed rate, air pressure, and size of openings for feed and air. The openings are varied by interchangeable parts of different sizes. In severe duties the liquid nozzle piece wears and has to be replaced periodically. With ceramics that are less abrasive, continuous operation for up to 6000 h has been recorded.

The cost of the compressed air is seldom important because of the high value added to the final piece. This is less true at high capacities because, in general, high-volume products have lower value. The weight ratio of compressed air to feed is 1.0 for two-fluid nozzles that

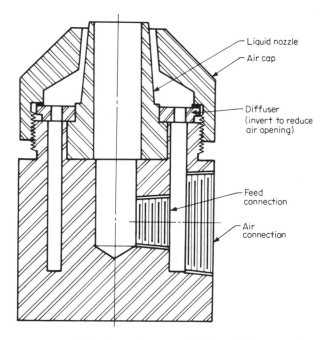

Figure 14.4 Abrasion-resistant two-fluid nozzle for ceramics.

mix air and feed externally. Internal mixing types, not often used on ceramics, have ratios of 0.5 to 0.6.

In spite of the wear, single-fluid nozzles, also called *hydraulic*, or *pressure* nozzles, are sometimes needed for the uniform drop size they produce, such as for isostatic pressing. Required pressures are 150 to 750 lb/in^2 (1030 to 5170 kPa) and the maintenance on nozzle and pump is relatively high, even though both have hard-surfaced wear parts.

Spinning discs. Some electronic ceramic manufacturing is at high capacities, such as certain ferrites and alumina for spark plugs. Large-diameter chambers and centrifugal atomizer wheels (spinning discs) are used. Proper wheel design is critical in considering weight against strength. The 6- to 10-in (15- to 25-cm)-diameter wheels turn at 8000 to 14,000 r/min and can fly apart if not well designed and constructed.

To make fine droplets using spinning discs, higher liquid velocities are required, which increases wear. One design combats abrasion on the principle of protecting inner surfaces by allowing them to become packed with the slip solids. The slip flows out of the wheel through tapered ceramic tubes. Another design conducts the slip in a smooth abrasion-resistant flow to a passage between flat ceramic inserts. The inserts are relatively thin, which results in a lighter wheel.

These designs run for many hours with little or no wear, but they are susceptible to thermal and mechanical shocks. They require careful handling and operating to avoid cracking. To protect the wheels, dryers are started up with a feed of hot water and not switched to the slip until equilibrium is reached. Proper care and pulling tools are used when removing a wheel from the drive shaft.

There are also wheel designs of lower cost. One has tungsten carbide wear pins that can be rotated for longer life. The pins separate two plates through which the slip flows, and they break up lumps and produce fine droplets. Another type produces drops of a narrow size range by the use of cylindrical passages for the slip. The passages are cast in an abrasion-resistant wheel of titanium, held in place by a stainless-steel band.

Heating and heat recovery. Natural gas is the fuel most often used for these dryers, and if there is no insulation, heat losses are high, especially from towers. With the outlet air temperature typically at about 300°F, heat recovery is cost-effective. The exhaust can heat the supply air to the burner using an air-to-air heat exchanger or the air-to-liquid (run-around) type.

Table 14.2 lists the functional design data for a typical tower spray dryer for a ceramic material. A standard 6-ft (1.8-m)-diameter unit with 1200-ft^3/min (0.566-m^3/s) airflow at the outlet should satisfy the

TABLE 14.2   Results for a Typical Ceramic Spray Dryer (U.S. Customary Units)

Miscellaneous data	
Feed temperature	60.0°F
Product temperature	178.3°F
Solid specific heat	0.400 Btu (lb · °F)
Elevation	0 ft
Elevation correction factor	1.000
Feed solids	65.00%
Product moisture	0.20%
Heat of crystallization	0.00 Btu/lb
Heat loss	2.09%
Calculated diameter	5.7 ft
Volume	399.9 ft^3
Volumetric factor	2.189
Cylinder height-to-diameter ratio	2.50
Cone angle	60.0°
Exposure	20.0 s
Auxiliary air	0.0%
Air leak	11.8%
Rates, lb/h	
Feed	464.5
Product	301.3
Solids	301.9
Evaporation	162.0
Product loss	1.2

Psychrometric data at 14.696 lb/in^2 (absolute)

Dry gas molecular weight	28.97
Specific heat at station 2	0.2434 Btu/(lb · °F)

Station	Temperature, °F	Moisture, lb/lb DA	Enthalpy, Btu/lb DA	Humid volume, ft^3/lb DA	Dry airflow, lb/min	ACFM, ft^3/min
1	60.0	0.00700	21.98	13.25	50.5	669
2	600.0	0.02240	175.99	27.68	50.5	1398
3	322.5	0.07586	169.14	22.13	50.5	1117
4	80.0	0.00200	21.35	13.65	6.0	81
5	300.0	0.06806	153.54	21.25	56.5	1200

Collector	Product, lb/h	Efficiency, %
1	242	80.0
2	59	98.0
3	0	0.0
Net	301	99.6

Heat and outlet air moisture data

Number of hydrogen atoms	4.11
Molecular weight	16.78
Higher heating value	23803 Btu/lb
Fuel moisture	0.0154 lb H_2O/lb DA
Saturation moisture	0.1095 lb H_2O/lb DA
Heat load	0.4666 MBtu/h
Dryer outlet (station 5)	
Adiabatic saturation ratio	62.2%
Relative humidity	2.2%
Dew-point temperature	114.5°F
Saturation temperature	129.6°F
Wet-bulb temperature	130.7°F

calculated 5.7-ft (1.7-m) diameter. Cylinder height for this size is 15 ft (4.6 m). The heat loss of 2.09 percent is typical for towers with an outlet air temperature of 300°F (149°C). The two-fluid nozzle introduces 6.0 lb/min (2.7 kg/min) of compressed ambient air, shown at station 4, giving a theoretical end of drying of 322°F (161°C) at station 3. This calculation method is conservative. It would be more accurate to have the compressed air reduce the inlet temperature.

Eighty percent of the production is taken out of the chamber as usable product. The heat source is a standard natural gas with an average of 4.11 hydrogen atoms per mole and a higher heating value of 23,803 Btu/lb (55,366 kJ/kg).

14.3   Rotary Drying of a Specialty Sugar*

A common application for direct rotary dryers involves food-grade sugars. In general, sugars require gentle thermal treatment because they tend to caramelize or char if overheated.

Traditionally, rotary dryers have been used on granulated sugar products for the following reasons.

1. Long retention times are possible at low temperature differentials.
2. Heating is uniform throughout the drying cylinder.
3. Attrition of the sugar's crystalline structure is minimal because of the gentle mechanical action.
4. Most other dryers cannot handle sugars because of their sticky nature.

Dryers for sugars are designed for cocurrent flow of air and product. The air is heated indirectly, using steam coils to comply with purity requirements for food-grade materials. The cocurrent arrangement takes advantage of the cooling effect of evaporation, making it possible for the wet feed to be exposed to the hottest air. In this way the rate of heat transfer is maximized without damage to the product.

A specialty sugar that had never been dried commercially was given an extended test program to determine its drying and handling characteristics. The intent was to establish the operating parameters for the design of a commercial rotary dryer and its system components.

Testing.   Prior testing and commercial experience on the drying of other sugars provided background for the test program. The tests were run in a standard rotary dryer test unit. Its major elements are shown

*D. W. Dahlstrom, ABB Raymond, and Edward M. Cook, Energy Saving Consultants.

in Fig. 14.5. Feed moisture was 24 percent and the material had to be dried down to a product moisture of 0.5 percent, with attrition controlled so that a minimum of fines would be generated. In addition, the final product was required to have a crystalline structure, acceptable for consumer use and suitable for normal packaging.

Tests proved that these specifications could be met. But they also showed that some of the material tended to agglomerate. The percentage was small enough that backmixing dry product into the wet feed was not needed. However, to get the wet material to shower properly in the hot air at the feed end, a radial, or unlipped, flight design was used. As the material dried farther along in the cylinder, it flowed more readily, and lifting and showering flights that were progressively cupped gave the desired performance. It was determined that the remaining small percentage of oversized agglomerates could be recycled upstream to the crystallizers.

Solids in a rotary dryer are transported by the rotation and slope of the cylinder and, with cocurrent flow, by the velocity pressure of the airflow on the showered particles. Unlike the more fully suspended-particle mode of flash, fluid-bed, and spray dryers, the particles in rotaries have longer contact with the vessel wall. In addition, the walls are hotter because the action is gentler and evaporation is not as rapid. The tests showed that the commercial unit for this sugar would require shell knockers to keep the wet material moving at the feed end.

Product samples were taken at various air and material temperatures, airflow rates, and solid retention times. These were tested for moisture content to help determine the best operating parameters for the commercial design. In addition, the active volume of the cylinder and the solids flow rate were measured. The ratio of these is the retention time; the loading is the volume of the cylinder occupied by the material. Testing demonstrated that good product was obtained at a retention time of just over 1 h and a loading of 15 percent.

The conditions that resulted in the best product were then used for a continuous 12-h test run. This confirmed the ability to produce continuously at the required quality with no buildup on the cylinder inside walls. During the test, particulate loading in the dryer's exhaust was also measured to aid in selecting and specifying the details of a fines collector. Furthermore, the extended test afforded the opportunity to observe any process difficulties not seen in short test runs.

Scaleup and design. Process data defined in the tests were used to scale up to the commercial unit. The size of the dryer was determined by the specified material flow rate and heat transfer considerations, as well as retention time, air and product temperatures, and air velocities from the test work.

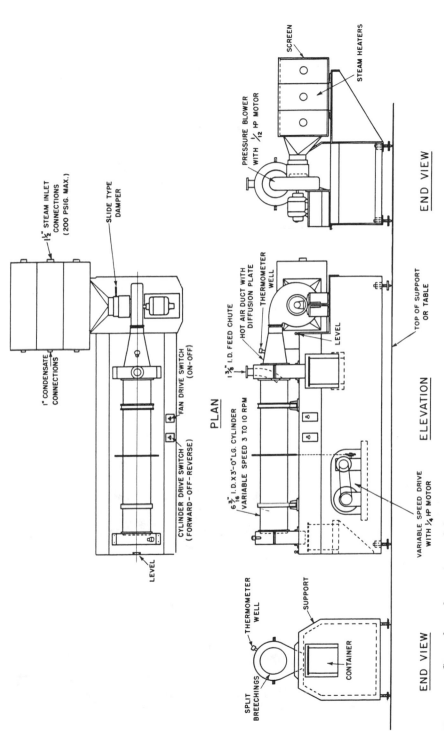

Figure 14.5 Steam-heated experimental dryer.

223

In addition, proper control of attrition and agglomeration was assured by using the same peripheral velocity of the test cylinder for the larger unit. This speed is also a factor in determining the cylinder's slope to give the correct retention time. A variable-speed drive was selected for a range of speeds, designed to bracket the test peripheral speed.

The most critical of the scaleup criteria are the heat transfer calculations. They help determine proper unit size and flight design. But prior experience, as well as fundamental heat transfer calculations, play a major role in this aspect of dryer design.

The commercial unit, shown on Fig. 14.6, was for the most part designed and built using preengineered rotary dryer components. Also incorporated were standard purchased items, such as a feed screw, steam-heated air heater, pulse-type bag filter, air locks, fans, dampers, and controls. A few parts had to be specially designed. These included the feed end flights, an integral trommel screen for scalping oversized agglomerated product at the discharge, and the corresponding discharge breeching to separate the oversize product from product that meets the specifications.

Commissioning and operation. After installation and before start-up some minor problems had to be resolved. The feed screw's mounting flange was reworked for proper fit to the inlet breeching, and its drive

Figure 14.6   Rotary dryer for specialty sugar.

motor was defective and had to be replaced. As is fairly common on dryer installations, there were a few electrical wiring problems to be corrected.

After initial settings and mechanical adjustments the system was first started cold without feed, and later run hot. At this stage the equipment was monitored carefully, and further adjustments were made. Next, process material was fed at about 20 percent of design rate, so that the initial run could be managed and observed easily.

After holding the reduced feed rate for 1 h, it was gradually increased to full capacity. To achieve the required retention time and product moisture level, final adjustments were made to the temperature controls, cylinder speed, and, by means of dampers, the airflow rate. In addition, the air pulsing rate on the bag filter collector was adjusted for optimum powder buildup on the fabric.

Air temperature controllers regulated the inlet air temperature and monitored the exhaust. A controller also monitored the product discharge temperature. These were the only controls needed to properly regulate all important variables and yield product at the quality required.

Table 14.3 lists the performance data and some of the physical data on the dryer and components. The heat loss of 4 percent of the heat load assumes that part of the shell is not insulated. The excess air, shown as "leak," is assumed to be at ambient conditions.

14.4 Fluid-Bed Drying of Amino Acids*

Fluid-bed processing enjoys wide use in the chemical, food, petroleum, petrochemical, mineral, pharmaceutical, and other industries for drying, heating, cooling, agglomerating, reacting, calcining, catalytic cracking, and combustion operations. Often the same unit is zoned to run two or more of these operations consecutively. The technique had its earliest beginnings about 1930 and has been practiced actively ever since.

Virtually all particulate substances can be fluidized, although there are some practical limits, such as particle size. A specific range of velocities is required for fluidization, governed by the particle density, size, shape, and other properties. A minimum velocity is needed, but if too high it elutriates an excess of fine material.

Fluid-bed dryers are designed for mixed flow, plug flow, and for combinations of both. The mixed flow is actually an advanced back-mixing operation because wet feed is introduced into a bed of nearly dry material. Turbulent fluidization provides the ideal action for con-

---

*John J. Walsh, Bepex Corporation.

TABLE 14.3   Data on Rotary Dryer for Specialty Sugar

Miscellaneous data	
Shell	
Diameter	2.0 ft
Length	20.0 ft
Rotation	6.0 r/min
Loading	15.0%
Slope	$\frac{1}{16}$ in/ft
Feeder	Screw conveyor
Heater	Steam heat exchanger
Drives	Three at 3 hp
Blowers	Two at 4 hp
Airflow	Cocurrent, 200 ft/min
Retention time	75 min
Feed and product characteristics	
Feed temperature	60.0°F
Product temperature	125.0°F
Solid specific heat	0.4 Btu/(lb · °F)
Feed moisture	24.0%
Product moisture	0.5%
Heat of crystallization	0.0 Btu/lb
Particle size	Granular grade, 70% + 200 mesh
Rates, lb/h	
Feed	250.0
Product	191.0
Solids	190.0
Evaporation	59.0

Psychrometric data at 14.696 lb/in^2	
Heat loss	4.0% (partly insulated)

Location	Tempera-ture, °F	Moisture, lb/lb DA	Enthalpy, Btu/lb DA	Humid volume, ft^3/lb DA	Dry airflow, lb/min	ACFM, ft^3/min
Heater inlet	60.0	0.00700	21.98	13.25	34.5	457
Dryer inlet	325.0	0.00700	86.72	20.01	34.5	690
Leak (estimated 26% of hot air)	60.0	0.00700	21.98	13.25	9.0	119
Dryer outlet	150.0	0.02965	69.43	16.11	43.4	700

Heat data and outlet air moisture data	
Heat source	Steam at 150 lb/in^2
Fuel moisture	0.0000 lb/lb DA
Saturation moisture	0.0414 lb/lb DA
Heat load	0.1339 MBtu/h
Dryer outlet data	
Adiabatic saturation ratio	71.6%
Relative humidity	18.0%
Dew-point temperature	88.7°F
Wet-bulb temperature	99.3°F

ditioning wet feed, bringing its moisture down rapidly and integrating it into the rest of the material.

Unique among the direct dryers, fluid beds can use indirect heaters for substantial energy savings and reduction of the system's volume. These heaters can be tube bundles or plate coils; the latter can also serve as baffles. Heat transfer rates to the solids are high because of the vigorous mixing action across the heat transfer surfaces.

Some of the important uses of fluid-bed dryers are for chemicals, foods, polymers, minerals, and pharmaceuticals. One of the most interesting is drying various amino acids, a family of pharmaceuticals that impose several drying problems. These materials need the ability of fluid beds to operate in both backmixed and plug flow modes, with their other traditional advantages, plus some innovations.

Amino acids are building blocks for body growth. They are used as food supplements in animal feeds, such as for beef and chicken. Hence they are dried and shipped in bags or in bulk to feed processors.

Amino acid feeds come to the dryer from a decanter centrifuge at 20 to 30 percent water content. They are lumpy and sticky. Final moisture content must be 0.5 to 1.0 percent. High heat input is needed at the start, where heat transfer dominates. Then long exposure time, or dwell time, is needed, because diffusion drying dominates in removing bound moisture.

Before discharge the product has to be cooled. Thus three zones are required, plus an initial zone to ensure getting the material well mixed and past the sticky phase. Different air velocities for the four zones provide the degree of fluidization and mixing required in each. A disengaging area above the bed reduces the elutriation of fines, which are captured in a cyclone collector and mixed with the product discharged from the last zone. Figure 14.7 is a diagram of the drying chamber.

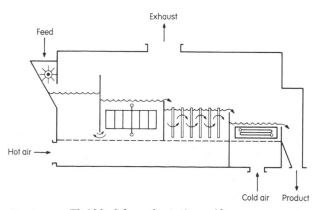

Figure 14.7   Fluid-bed dryer for amino acids.

Zone 1.   The first zone has a deep bed and good mixing action to blend wet feed into a *sink* of dry bed material—a kind of blotter effect. A special pin-mill feeder broadcasts the feed across the top of the bed, so that it disperses uniformly through the 4-ft (1.2-m) depth of relatively dry material. The velocity of the 300°F (149°C) air is 160 ft/min (0.8 m/s) to fluidize, mix intensively, and break up lumps. Solids could build up, and even bake onto hot bare metal surfaces, so an internal heat exchanger cannot be used. Only part of the heat is supplied by the hot air. The balance is supplied by solids that flow back from the second zone under an underflow baffle, which separates the bed inventories of the first and second zones.

Zone 2.   In the second zone, as in the first, heat transfer dominates. In this zone the moisture content of the particles is below the sticky phase, so an internal heat exchanger can be used to provide much of the heat duty. Its tubular design does not interfere with good backmixing and allows the added heat to be carried by solids back to the first zone.

Between these two zones the underflow baffle is 6 ft (1.8 m) high, but with a 1-ft (0.3-m) clearance at the bottom, allowing solids to pass under it in both directions—another form of backmixing. The bed height is set by the 4-ft (1.2-m)-high weir, over which the solids flow into the third zone.

Zone 3.   The high length-to-width ratio of this zone is designed for plug flow. Particles entering have mostly bound moisture and need a long exposure time to allow the remaining 3 to 5 percent moisture to diffuse out. Heat exchanger coils are the plate type, arranged in a zigzag pattern to serve as baffles and minimize backmixing. The inlet air at 300°F (149°C) has a velocity of only 120 ft/min (0.6 m/s) to minimize solids carryover in the outlet airstream in this zone.

Zone 4.   Because the particles have to be cooled to 104°F (40°C) before discharge, the last zone has mixed flow with cooled, conditioned air at 77°F (25°C) at a velocity of 120 ft/min (0.6 m/s). Water-cooled plate coils provide an added cooling effect, and the product is discharged over an overflow weir through a rotary air lock into bagging equipment.

Miscellaneous design features.   The screen supporting the particles has canopied openings that prevent backsifting of material into the inlet air plenum. By directing the incoming air parallel to the plane of the screen, heavier (thus difficult to fluidize) particles are directed toward the discharge point. There they are removed periodically by means of a manually operated underflow dump gate. This gate is also used to empty the bed inventory prior to long-term scheduled shutdown. There is easy access to the screen and internal heating elements, which may require manual washing to be cleaned after operational upsets.

The system has three blowers—for the drying zones, the cooling zone, and the exhaust. The velocity to each of the zones is set by manually operated dampers. A pneumatically actuated damper on the exhaust fan controls the null point pressure in the freeboard of the fluid-bed chamber.

Testing. When amino acids were first tested, the normal test program was followed, using a batch-fluidized bed column. Data were taken of air velocity, pressure drop, and fines carryover to establish acceptable fluidizing conditions. Moistures and temperatures were recorded against time to get data for the drying curve and exposure time requirements.

Subsequently pilot-scale continuous tests were conducted to determine the heat transfer coefficients for the internal heat exchangers in each zone. These tests also confirmed fluidizing velocities, solids carryover in the outlet airstream, exposure time, and the bed moisture and temperature conditions for all zones.

Table 14.4 lists some of the major design conditions for a four-zone fluid-bed drying system for amino acids. Air at 300°F (149°C) flows through each of the three heated zones at differing velocities, and cooled air flows through the last zone. Internal heat exchangers in all but the first zone greatly reduce the energy requirements.

TABLE 14.4 Typical Operating Data for Amino Acids

Feed temperature	70°F
Feed moisture	20–30%
Product temperature	104–110°F
Product moisture	0.5–1.0%

	Zone			
	1	2	3	4
Operation	Constant-rate drying	Constant-rate drying	Falling-rate drying	Cooling
Weir or baffle height, in	72	48	42	36
Internal exchanger	None	Vertical tubes	Zigzag plate	Parallel plate
Heating/cooling medium		Steam	Steam	Chilled water
Fluidizing velocity, ft/min	160	140	120	120
Inlet air temperature, °F	300	300	300	77
Bed temperature, °F	165	200	250	104
Bed moisture, %	5.0	3.5	0.5	0.5
Bed area, ft²	20	30	30	20
Solids dwell time, min	20*	30*	40	20

*Zones 1 and 2 are not residence-time controlled, but the weir height in zone 2 (required to cover the internal heat exchanger) results in a deep bed inventory and long residence time.

A

Method for
Conversion of Units

Table A.1 lists conversion factors for making changes between units. Table A.2 is for changes in magnitude and the prefixes for common magnitudes. In Table A.1 the SI base unit for each category is listed first. Four of the base units are in kilo--kg, kPa, kJ, and kW. This choice of units gives the most practical values for drying and some related operations.

The simplest conversion is to change to a base unit. Multiply the given amount by the factor given in Table A.1. Thus, changing 1.2 lb to kg,

$$1.2 \text{ lb} \times 0.45359 = 0.54431 \text{ kg}$$

Two other possible operations are to change magnitudes and to build conversion factors for combination units. For both, a simple systematic method is to insert each pair of values in the ratio of FROM/TO. For example, to change from kg to mg, the FROM value (kilo) is $1E+3$, the TO value (milli) is $1E-3$. The FROM/TO ratio is $1E+3/1E-3$, which gives $1E+6$. ($1E-n$ represents 10^{-n}.) Changing the example from kg to mg then becomes

$$0.54431 \text{ kg} \times 1E+6 = 544{,}310 \text{ mg}$$

If the change of magnitude is for units with an exponent other than 1, the FROM/TO ratio must be raised to that exponent. Thus changing from dm^3 (L) to cm^3, the ratio is $1E-1/1E-2$ to the third power, or $1E+3$.

TABLE A.1   Conversion Factors to and from U.S. Customary and SI Units

Unit	Abbreviation	Factor*
Length		
meter	m	1
angstrom	Å	1 E–10†
micron	μm	1 E–6
mil (0.001 in)	mil	2.5400 E–5
centimeter	cm	0.01
inch	in	0.02540
foot	ft	0.30480
yard	yd	0.91440
mile	mi	1609.3
Area		
square meter	m^2	1
square centimeter	cm^2	1 E–4
square inch	in^2	6.4516 E–4
square foot	ft^2	0.092903
acre	acre	4046.9
hectare	ha	10,000
Volume, gas, and liquid		
cubic meter (kL)	m^3	1
cubic centimeter (cc)	cm^3	1 E–6
ounce (liquid)	oz	2.9574 E–5
cubic inch	in^3	1.6387 E–5
liter (dm^3)	L	1 E–3
US gallon	gal	3.7854 E–3
UK gallon	gal	4.5461 E–3
cubic foot	ft^3	0.028317
barrel (42 gal)	bbl	0.15899
Mass		
kilogram	kg	1
grain	gr	6.4799 E–5
ounce	oz	0.028350
pound	lb	0.45359
short ton (2000 lb)	ton	907.18
metric ton	t	1000 (tonne)
long ton (2240 lb)		1016.0
Mass, air		
kilogram	kg	1
standard cubic foot, 60°F	scf	0.034630
normal cubic meter, 0°C	norm. m^3	1.2930
Pressure, absolute		
kilopascal	kPa	1 (Pa = N/m)
kilogram/square meter	kg/m^2	9.8068 E–3
torr, 0°C	mm Hg	0.13332
inch WG, 60°F	in WG	0.24884
foot WG, 60°F	ft WG	2.9861
inch Hg, 60°F	in Hg	3.3864
pound/square inch (psia)	lb/in^2	6.8948
bar	bar	100
atmosphere	atm	101.33

TABLE A.1   Conversion Factors to and from U.S. Customary and SI Units (*Continued*)

Unit	Abbreviation	Factor*
Heat, energy, work		
kilojoule	kJ	1
foot-pound	ft · lb	1.3558 E − 3
Btu (IST)	Btu	1.0551
kilocalorie (IST)	kcal	4.1868
horsepower-hour	hp · h	2684.5
kilowatt-hour	kW · h	3600
Power		
kilowatt	kW	1
foot-pound/min	ft · lb/min	2.2597 E − 5
Btu/hour	Btu/h	2.9307 E − 4
calorie/second	cal/s	4.1868 E − 3
HP (electric)	hp	0.746
tons refr.	12,000 Btu/h	3.5169
Time		
second	s	1
minute	min	60
hour	h	3,600
day	d	86,400
year (365 d)	yr	31,536,000
Temperature difference		
centigrade, kelvin	°C, K	1
Fahrenheit, Rankine	°F, °R	0.55556

Common Combinations		
From	To	Factor
ft^3/min (cfm)	m^3/s	4.7195 E − 4
ft/min	m/s	5.0800 E − 3
ft^3/lb	m^3/kg	6.2429 E − 2
centipoise (cP)	kg/(m · s)	1 E − 3
lb/(ft · s)	kg/(m · s)	1.4882
Btu/lb	kJ/kg	2.3260
Btu/(h · ft^2)	W/m^2	3.1546
Btu/(lb · °F)	kJ/(kg · K)	4.1868
Btu/(h · ft^2 · °F)	W/(m^2 · K)	5.6783
gr/ft^3	mg/m^3	2288.3
gr/std. ft^3, 60°F	mg/norm. m^3, 0°C	2419.4

*Values are given to five significant digits. Trailing zeros are omitted from exact values. Force designation is omitted.

†E − n represents 10^{-n}.

TABLE A.2   Common Magnitude Prefixes

Prefix	Symbol	Conversion factor*
pico	p	1E−12
nano	n	1E−9
micro	μ	1E−6
milli	m	1E−3
centi	c	1E−2
deci	d	1E−1
One	—	1
deka	da	1E+1
hecto	h	1E+2
kilo	k	1E+3
mega	M	1E+6
giga	G	1E+9
tera	T	1E+12

*1E−n represents 10^{-n}.

A conversion factor for a combination of units can best be shown by example: Btu/(min · ft^2 · °F) to kW · h/(h · m^2 · °C). Set the FROM/TO pair Btu/(kW · h) over the three FROM/TO pairs for the times, areas, and temperature differences. Converting

$$\frac{\text{Btu}}{\text{min} \cdot \text{ft}^2 \cdot \text{°F}} \text{ into } \frac{\text{k} \cdot \text{Wh}}{\text{h} \cdot \text{m}^2 \cdot \text{°C}},$$

We have

$$\frac{\text{Btu TO k} \cdot \text{Wh}}{(\text{min TO h})(\text{ft}^2 \text{ TO m}^2)(\text{°F TO °C})} = \frac{1.0550/3600}{(60/3600)(0.092903/1)(0.55556/1)}$$

$$= 0.34068$$

which is the conversion factor.

B

Miscellaneous Data

Improved data values become available from time to time; thus the information presented here may not be the most recent. When assembled, it was the best available for the purpose and sufficiently accurate. Various sources were used in gathering the data, but most data were provided by the references listed in the Bibliography.

TABLE B.1 Air Data

a. Normal Composition of Dry Air			
Gas	Vol %	Wt %	Molecular weight
Nitrogen	78.09	75.52	28.016
Argon	0.93	1.28	39.948
Carbon dioxide	0.03	0.04	44.010
Inert subtotal/average	79.05	76.84	28.162(average)
Oxygen	20.95	23.15	32.000
Total/average	100.00	99.99	28.966(average)

b. Miscellaneous Air Data	
1.0 atmosphere	14.696 lb/in^2; 101.33 kPa; 33.93 ft WG; 29.92 in Hg; 760 mm Hg
1.0 lb/in^2	27.708 in WG; 2.3090 ft WG; 6.8948 kPa
Standard air density	0.075 lb/ft^3; 1.201 kg/m^3
Specific volume [dry air at 1.0 atm and 70°F (21.1°C)]	13.33 ft^3/lb; 0.8322 m^3/kg
Air density correction	0.964 at 1000-ft (305-m) elevation; 0.832 at 5000 ft (1524 m)

For additional data on air pressure versus elevation, design temperatures for various locations, and other information useful for dryer design see Jorgensen (1983).

c. Temperature Data														
°C	−40	0	10	20	30	40	60	80	100	150	200	300	400	500	600
°F	−40	32	50	68	86	104	140	176	212	302	392	572	752	932	1112

0 K = 0°R; −273.16°C; −459.69°F

Temperature differences: 1°C = 1.8°F; 10°C = 18°F; 50°C = 90°F; 100°C = 180°F

TABLE B.2 Enthalpies of Dry Air and Water Vapor

| Temperature | | Dry air | | Water vapor |
°R	°F	Enthalpy,* Btu/lb	Specific heat, Btu/(lb · °F)	enthalpy,‡ Btu/(lb · °F)
460	0	109.90	0.2396†	
510	50			1082.6†
560	100	133.86	0.2396	1105.0
660	200	157.92	0.2401	1150.1
760	300	182.08	0.2405	1195.7
860	400	206.46	0.2414	1241.8
960	500	231.02	0.2422	1288.5
1060	600	255.96	0.2434	1336.7
1160	700	281.14	0.2446	1384.5
1260	800	306.65	0.2458	1433.7
1360	900	332.48	0.2472	1483.8
1460	1000	358.63	0.2487	1534.8
1560	1100	385.08	0.2501	1586.8
1660	1200	411.82	0.2515	1639.6
1760	1300	438.83	0.2530	1693.4
1860	1400	466.12	0.2544	1748.1
1960	1500	493.64	0.2558	1804.3†
2460	2000	634.34	0.2622	2095.6
2860	2400			2343.5
2960	2500	778.97	0.2676	

*For values of K_j [Eq. (5.27)] subtract 109.90 from dry-air enthalpy to convert to 0°F basis.
†Extrapolated values.
‡Superheated water vapor enthalpies are at 1.0 lb/in² absolute. Partial pressure of water vapor in a dryer may be as high as 4.0 lb/in², but even at that pressure, enthalpies are at most only 0.04% less.

TABLE B.3 Data for Common Fuels

Fuel	Number of hydrogen atoms N_h	Molecular weight M_h	Gross heating value H_v, Btu/lb	Slope S_l§
Methane	4.0	16.04	23,885	2.72×10^{-5}
Natural gas*	4.105	16.78	23,803	2.67×10^{-5}
Natural gas†	4.300	18.14	23,651	2.64×10^{-5}
Ethane	6.0	30.07	22,323	2.28×10^{-5}
Propane	8.0	44.09	21,560	2.13×10^{-5}
No. 2 fuel oil‡	13.73	110.0	19,100	1.61×10^{-5}

*90% methane, 5% ethane, 5% nitrogen.
†85% methane, 15% ethane.
‡Values estimated from carbon-hydrogen ratio of 6.93.
§Slope of the line of combustion moisture (y axis) plotted against heater temperature difference: outlet minus inlet (x axis). Slope times heater temperature difference gives an estimate of moisture added by combustion.

TABLE B.4   Heat Loss through Insulation*

Hot face temperature, °F	Cold face temperature, °F	Heat loss, Btu/(h · ft²)	Heat transfer rate, Btu/(h · ft² · °F)
150	85.3	7.2	0.103
200	89.2	13.2	0.110
250	93.3	19.8	0.116
300	97.7	27.1	0.123
350	102.3	35.2	0.130
400	107.2	44.1	0.138
450	112.4	53.9	0.146
500	117.9	64.6	0.154
600	129.7	88.9	0.171
700	142.8	117.5	0.190
800	157.1	150.8	0.209
900	172.6	189.2	0.231
1000	189.3	233.1	0.253
1100	207.1	283.1	0.278
1200	225.9	339.5	0.303
1300	245.7	402.9	0.330
1400	266.3	473.5	0.359

*Values are based on Eagle-Picher data for 2-in mineral fiber blanket insulation on a vertical surface and at an ambient temperature of 80°F (27°C). For 3-in insulation thickness, heat transfer rates are 68% less. From horizontal surfaces the rates would be somewhat lower.

TABLE B.5   Sieve Sizes

Opening, μm	Mesh number	Opening, μm	Mesh number
44	325	175	80
53	270	250	60
62	230	300	50
74	200	350	45
88	170	420	40
105	140	590	30
125	120	840	20
150	100		

TABLE B.6 Solvent Properties

Solvent	Molecular weight M_v	Boiling point T_b, °F	Latent heat of vaporization L_h, Btu/lb	Liquid specific heat C_q, Btu/(lb·°F)	Coefficients for vapor specific heat, Eq. (13.30)			
					Z_a	$Z_b \cdot 10^2$	$Z_c \cdot 10^6$	$Z_d \cdot 10^9$
1 Acetone	58.08	133.7	223.9	0.5337	1.6250	6.6610	−37.3700	8.3070
2 Benzene	78.11	176.2	169.5	0.4208	−9.0726	11.7135	−76.7622	18.9559
3 Carbon tetrachloride	153.84	170.2	83.5	0.2126	11.9687	3.5274	−31.5153	9.3877
4 Chloroform	119.39	143.1	103.2	0.2300	7.2295	3.6084	−28.3455	7.9196
5 Ethanol	46.07	173.1	367.9	0.6170	2.4887	5.0042	−19.7006	0.9396
6 Heptane	100.20	209.5	137.5	0.5357	−1.2293	16.1454	−87.2008	18.2892
7 Hexane	86.17	156.2	144.2	0.5459	−1.0543	13.8990	−74.4861	15.5057
8 Methanol	32.04	148.5	472.9	0.6073	5.9399	1.2151	14.0030	−10.7783
9 Methyl ethyl ketone	72.10	175.3	190.6	0.5351	7.3460	0.5760	0.0000	0.0000
10 Methylene chloride	84.94	103.5	140.8	0.2763	4.0387	3.3547	−22.4582	5.6823
11 n-Propanol	60.09	208.0	295.4	0.6220	3.4948	6.4656	−20.8853	−1.4179
12 i-Propanol	60.09	180.5	286.0	0.6958	−0.3913	8.6991	−51.6641	11.8272
13 Toluene	92.13	231.4	156.2	0.4289	−8.4058	13.4513	−83.5474	19.7259
14 p-Xylene	106.00	281.3	146.3	0.4206	−6.1462	14.5476	−83.4712	18.2531
15 Water	18.02	212.0	970.3	1.0000	7.3000	0.2460	0.0000	0.0000

TABLE B.7   Explosive Limits for Some Common Solvents

	Molecular weight	Specific gravity	LEL*	UEL*
Acetic acid	60	1.05	5.4	—
Acetone	58	0.79	2.1	13.0
Amyl alcohol	88	0.81	1.2	—
Benzene	78	0.88	1.4	7.1
Butanol (tertiary)	74	0.79	2.3	8.0
Butyl acetate	116	0.88	1.4	7.6
Butyl alcohol	74	0.81	1.4	11.6
Butylene	56	—	1.7	9.0
Carbon disulfide	76	1.26	1.0	50.0
Cellosolve	90	0.93	2.6	15.7
Chlorobenzene	113	1.11	1.8†	9.6‡
Decane	142	0.73	0.8	5.4
Dichlorobutane	125	—	1.5	4.5
Dimethyl formamide	73	0.95	2.2†	—
Dodecane	170	0.75	0.6	—
Ethyl alcohol	46	0.79	3.5	19.0
Ethyl ether	129	0.71	1.9	36.5
Ethylbenzene	106	0.87	1.0	—
Ethylene dichloride	99	1.26	6.2	15.9
Heptane	100	0.69	1.2	6.7
Hexane	86	0.66	1.2	7.5
Kerosene	276	0.81	0.6	5.6
Methyl alcohol	32	0.79	5.5	36.5
Methyl chloride	50	0.92	7.6	19.0
Methyl ethyl ether	60	0.75	2.0	10.1
Methyl ethyl ketone	72	0.81	1.8	10.0
Methylene chloride	85	1.34	15.5§	66.9§
Octane	114	0.71	0.8	3.2
Pentane	72	0.63	1.4	8.0
Propyl alcohol (iso)	60	0.79	2.5	12.0
Propyl alcohol (n)	60	0.80	2.0	12.0
Toluene	92	0.87	1.3	7.0
Vinyl chloride	63	0.97	4.0	22.0
Xylene (para)	106	0.87	1.0	7.0

*Values of LEL and UEL were accumulated over several years, but some sources differ, possibly because of differing test temperatures or methods.
†At 212°F.
‡At 309°F.
§In oxygen.

Glossary

The following definitions are those that are peculiar to drying, or that cannot be easily found using the index. The most extensive explanations, however, generally appear in the text.

ACFM   Volumetric flow rate, often expressed as actual cubic feet per minute.

Adiabatic   Without change of heat (**enthalpy**, or **heat content**, remains constant). Although heat enters a direct dryer continuously in the air and feed, there is no separate source of heat added to the system as to an indirect, or **nonadiabatic** dryer.

Adiabatic saturation ratio (ASR)   Ratio of air moisture divided by saturated air moisture at the same enthalpy in lb/lb (kg/kg). It is usually expressed as a percent, even though called a ratio.

Air   Combined dry gas and vapor with it; also a drying term for fired heater exhaust, which is a reduced oxygen mixture of combustion gases plus dilution air.

Air moisture   Weight ratio of vapor to dry air in lb/lb (kg/kg). The term **absolute humidity** is also commonly used in engineering practice.

Ambient   Air surrounding equipment, such as the heater.

Angle of repose   Measurement of a solid's ability to flow. It is the angle that particles form with the horizontal when poured slowly onto a flat surface; a smaller angle for freer flow.

Backmixing   Incorporating dry product into wet solids before or at the start of drying to improve feed consistency. In a fluid-bed dryer, it is the addition of wet feed into the bed.

Bed   Mass of particulates in some dryers, at rest or stirred by an agitator or by airflow. In some dryers, such as rotaries, part of the bed of material is picked up and showered down to improve mixing and moisture removal.

Bound moisture   Liquid held in a solid by physical or chemical means, and thus not easily removed. It exerts a vapor pressure less than that of free liquid at the same temperature.

Capillary flow   Liquid flow through small openings in a solid or across its surface, caused by surface tension.

Constant-rate period   Period of supplying enough liquid to surface to maintain a constant rate of evaporation.

Critical moisture content   Moisture of solid at the end of constant-rate and start of falling-rate periods.

Density   Property of products measured in a variety of ways. Bulk density is the most common, for which a known volume is weighed to determine the **as poured**, or **loose**, density. A **settled, tapped**, or **packed** density is found by tapping the container. **Particle** density and **absolute** density are less commonly required. The usual units are lb/ft^3 (g/cm^3 or kg/m^3).

Dew point   Temperature at which vapor begins to condense from a gas-vapor mixture that is being cooled.

Diffusion dominated drying   Drying characterized by moisture that is trapped firmly and requires a long exposure time to be released. Heat transfer rate is low and in the falling-rate period.

Direct drying   Heating medium comes into direct contact with wet solids. It is said to be **adiabatic** because (if perfectly insulated) there is no gain or loss of heat in the system. It is also called **convection** drying because the heat moves largely by convection.

Direct-fired heater   Adds heat to drying air by combining products of combustion with dilution air.

Dry air (DA) or other dry gas (DG)   Nonvapor portion of drying medium that cannot condense under existing conditions. Air (or low-oxygen air) is the drying medium for most direct dryers. Nitrogen is used for most closed-cycle duties; other gases are used rarely.

Drying   In this text, evaporating moisture by heating a wet material, leaving a relatively dry solid.

Drying chamber   One of many terms used for housing of a drying vessel. Others include **cylinder, shell, trough, tower**, and **duct** or **tube** (for flash dryers). The choice depends largely on the dryer type and geometry.

Dust loading   Amount of powder in air or gas stream, expressed in gr/ft^3 (grains; mg/m^3).

Elutriation   Removal of fine powder by a gas stream.

Enthalpy   Heat content in Btu/lb (kJ/kg). For a mixture it is the sum of the individual enthalpies. Enthalpy is measured above a reference temperature, which in U.S. practice is 0°F for air and 32°F (0°C) for water vapor. For nonaqueous operations and in all metric systems the basis is 0°C for both gas and vapor.

Exposure (or residence, retention, or dwell) time   Time spent in dryer by solids. It may be expressed as an average or as a range of time distribution. If the solid's time cannot be measured or calculated, the time the air spends in the dryer is calculated. The units may be seconds, minutes, or hours, depending on the application and type of dryer.

Falling-rate period Period when drying rate is no longer constant, but falls off continuously.

Feed, product, and material Solid and its moisture before, after, and during drying. The term **solid** refers to dry solid or, when the context is clear, to the solid and its moisture.

Flow (of air or other gas) Given either by volume (such as ft^3/min or m^3/min) or by weight (such as lb/h or kg/h). In relation to solids in the airstream in a dryer, the flow is most often **cocurrent**, less often **countercurrent**. In **mixed flow** the flow changes from one mode to the other. **Parallel flow** is cocurrent, but sometimes implies a flow without turbulence.

Fluid Any liquid, vapor, gas, or their mixtures. Terms **fluid bed** and **fluidized** refer to particles activated into a fluidlike condition by forcing a gas up through them.

Heat capacity See **specific heat**.

Heat content See **enthalpy**.

Heat loss Heat escaping a system from vessel surface, in Btu/h (kJ/h). Although usually referred to as radiant heat loss, it is actually a combination of radiant and convective heat transfer. Heat can also be considered as lost in the air and solids leaving the system warmer than they entered. Some of this loss can be recovered.

Heat of crystallization Heat either added to or subtracted from a solid when it crystallizes. The amount, in Btu/lb (kJ/kg), is equal in value but opposite in sign to heat of solution, which is listed in the literature, usually in kg · cal/g · mol.

Heat source Means by which heat is applied to dryer—condensing steam, burning fuel, electricity, thermal fluid, or other.

Heat transfer dominated drying Drying characterized by moisture at the surface of a material that evaporates readily at a constant rate at a high input of heat.

Heat transfer rate Heat exchanged from one material to another at a rate of Btu/(h · ft^2 · °F) [W/(m^2 · K)].

Humidity or, more accurately, absolute humidity Weight of vapor per unit weight of dry gas in lb/lb DA (kg/kg DA). See also **relative humidity** and **adiabatic saturation ratio**.

Humid volume Volume of dry air (or other gas) and its vapor per unit weight of dry air only in ft^3/lb DA (m^3/kg DA).

Indirect drying Using a heating medium (usually steam) that does not contact the solids. It is said to be **nonadiabatic** because heat is added from outside the system. It is also called **conduction drying** because most of the heat reaches the solids by conduction through a metal wall.

Indirect heater   Adds heat to the air or other gas drying medium by use of a heat exchanger. Its exhaust does not enter the dryer.

Inert   Refers to nitrogen or other less reactive gases, but is also used for air with reduced oxygen content. A dryer is said to be **self-inertized** when it uses combustion gases and recycled dryer exhaust to reduce the air's oxygen content.

Instantized   Refers to a product that has been made easier to disperse in a liquid, usually water. Agglomeration is one method of giving fine powders this quality.

Latent heat of vaporization   Heat needed to vaporize a unit weight of liquid, in Btu/lb (kJ/kg).

Loading, fill, coverage, hold-up   Extent to which a drying vessel is filled with material; expressed as a percent of either the vessel's total volume or coverage of the agitator. See also **dust loading**.

Moisture   Refers to either liquid or vapor (water and water vapor in most applications). As liquid it may be in feed or product and is expressed as a percent wet basis. As vapor it is in air (or other gas) and is most commonly expressed as lb water vapor/lb DA (kg/kg DA).

Recycle   Product returned for further drying, or sent to either the dryer or a backmixer to improve the consistency of the wet feed. The term is also applied to the return of part or all of the drying gas to the dryer, after some moisture has been removed, usually by condensation.

Relative humidity   Partial pressure of water vapor in air over vapor pressure of liquid water at same temperature, times 100 to express it as a percent. For other liquids the term **percent saturation** is used. See also **adiabatic saturation ratio**.

Scaleup   Performance of a small unit and how it relates to a larger one, usually for design.

Sensible heat   Heat needed (with no evaporation) to raise the temperature of a unit weight of material 1.0 degree, expressed as Btu/(lb · °F) [kJ/(kg · K)].

Solvent   Any nonaqueous liquid (nearly all are organic) and often used even if the solids are not soluble in it. This helps to set these liquids apart from water, which, although the most common solvent, is classified separately from them in drying terminology.

Specific heat or heat capacity   Heat required to raise 1.0 lb of substance 1.0°F at constant pressure, in Btu/(lb · °F) [kJ/(kg · K)].

Specific volume   Reciprocal of density, in ft^3/lb (m^3/kg). Term is sometimes used when **humid volume** is meant.

Stickiness or tackiness   Tendency of a material to stick to surfaces or to itself. If not properly controlled, it hinders breaking down a feed into discrete particles, slows drying, and causes fouling of surfaces.

Supply air   Air that enters heater.

Temperature   Dry-bulb temperature unless otherwise specified. See also **wet-bulb temperature**.

Thixotropic   Refers to pastes that become less viscous when subjected to shear forces, such as by stirring and pumping. If made thin enough, they can be sprayed. But the process is reversible, and viscosity increases again when the shear forces stop.

Time   See **exposure time**.

Turndown   Degree to which equipment, particularly burners, can be reduced in capacity.

Vapor   Component of a gas that may partially condense under the existing conditions.

Wet-bulb temperature   Temperature at equilibrium between heat and mass transfer in which liquid evaporates from a small mass into a mass large enough to be unaffected by the added moisture. As long as the wick of a wet-bulb thermometer and the velocity of airflow past it maintain the equilibrium, the temperature remains constant, lower than the dry-bulb. The extent of the cooling effect depends on the difference between the moisture in the air and the moisture at saturation.

Bibliography

A.I.Ch.E.: "Spray Dryer Testing Procedure," Am. Inst. of Chemical Engineers, New York, 1988.

Cook, E. M., and H. D. DuMont: "New Ideas to Improve Dryer Performance," *Chem. Eng.*, May 1988.

———— and R. W. Lang: "Optimum Design of Multi-Stage Drying Systems," *Chem. Eng. Prog.*, Apr. 1979.

Dittman, F. W.: "How to Classify a Drying Process," *Chem. Eng.*, pp. 106–108, Jan. 17, 1977.

Jorgensen, R. (ed.): *Fan Engineering*, 8th ed., Buffalo Forge Co., Buffalo, N.Y., 1983.

Keenan, J. H., and J. Kaye: *Gas Tables*, Wiley, New York, 1948.

————, F. G. Keyes, P. G. Hill, and J. G. Moore: *Steam Tables*, Wiley, New York, 1978.

Keey, R. B.: *Introduction to Industrial Drying Operations*, Pergamon, Elmsford, N.Y., 1978.

Marshall, W. R.: *Atomization and Spray Drying*, Chem. Eng. Prog. Monograph ser. 50, 1954.

Masters, K.: *Spray Drying*, Wiley, New York, 1976.

McAdams, W. H.: *Heat Transmission*, McGraw-Hill, New York, 1954.

Mollier, R.: "Ein neues Diagram für Dampfluftgemische," *Z. VDI*, vol. 67, pp. 869–872, 1923.

Mujumdar, A. S. (ed.): *Handbook of Industrial Drying*, Marcel Dekker, New York, 1987.

Perry, R. H., and D. W. Green: *Perry's Chemical Engineers' Handbook*, 6th ed., McGraw-Hill, New York, 1984.

Reay, D.: "Theory in the Design of Dryers," *The Chem. Eng.*, pp. 501–503, 506, July 1979.

Root, W. L.: "Indirect Drying of Solids," *Chem. Eng.*, pp. 52–64, May 2, 1983.

———— and E. M. Cook: "Energy Use in Paddle Dryers," *Chem. Eng. Prog.*, Mar. 1981.

Scarrah, W. P.: "Improve Production Efficiency via Evolutionary Operation," *Chem. Eng.*, Dec. 7, 1987.

Thinh, T. P., et al.: "Equations Improve C_p^* Predictions," *Hydrocarbon Process*, pp. 98–104, Jan. 1971.

Treybal, R. E.: *Mass Transfer Operations*, McGraw-Hill, New York, 1980.

U.S. Dept. of Commerce: "Monthly Normal Temperatures, Precipitation, and Degree Days," Tech. Paper 31, 1956.

Index

ABOUT THE AUTHORS

EDWARD M. COOK is a drying systems consultant, specializing in energy-saving strategies. Over a period of 30 years in the drying industry he developed calculation methods and computer programs for spray, flash, and fluid-bed dryers; for energy conservation in dryers; and for psychrometric charts. Before that he spent 13 years in the heat-exchanger industry. He has a B.Ch.E. degree from Polytechnic Institute of Brooklyn, is a member of AIChE and ASME, and has authored over 40 articles on dryers and heat exchangers.

HARMAN D. DUMONT has worked in the drying industry for over 40 years. He founded DuMont Drying Consultants, Inc., in 1978 to serve makers and users of dryers and other process equipment. Previously he held positions as project engineer and manager of test and development at a major dryer firm. DuMont has managed the testing of hundreds of materials on spray, flash, and fluid-bed dryers and has been the principal engineer in the design and start-up of over 50 drying systems all over the world. He has a B.S. degree in management and an A.A. degree in mechanical engineering from Rutgers University, is a member of AIChE, and has coauthored several articles on dryers.